CHEMISTRY RESEARCH AND APPLICATIONS

EDTA

SYNTHESIS, USES AND ENVIRONMENTAL CONCERNS

CHEMISTRY RESEARCH AND APPLICATIONS

Additional books in this series can be found on Nova's website under the Series tab.

Additional e-books in this series can be found on Nova's website under the e-book tab.

CHEMISTRY RESEARCH AND APPLICATIONS

EDTA

SYNTHESIS, USES AND ENVIRONMENTAL CONCERNS

ANDRIS MOLNAR
EDITOR

nova publishers
New York

Additional color graphics may be available in the e-book version of this book.

Library of Congress Cataloging-in-Publication Data

ISBN: 978-1-62808-146-6

Published by Nova Science Publishers, Inc. † New York

CONTENTS

Preface		**vii**
Chapter 1	Symbiotic and Synergic Effects in Amide and Ester Derivatives of EDTA *Francisco Meijide, Juan V. Trillo, Santiago de Frutos, Aida Jover, José Vázquez Tato, Victor H. Soto and Luciano Galantini*	**1**
Chapter 2	Structural and Electrical Properties of $CaCu_3Ti_4O_{12}$ Obtained from Different EDTA-Gel Derived Powders *Filipe Amaral, Manuel Valente, Manuel Pedro Graça and Luís Cadillon Costa*	**45**
Chapter 3	Photo- or Radiation-Induced Degradation of EDTA and Its Complexes of Heavy Metals *Katerina Cubova and Barbora Drtinova*	**91**
Chapter 4	The Use of EDTA Root Conditioning in Regenerative Procedures of Periodontal Tissues *Daniela Leal Zandim-Barcelos and José Eduardo Cezar Sampaio*	**119**
Chapter 5	Effects of EDTA on Adsorption Behavior of Cadmium onto Natural Freshwater Surface Coatings *Xiuyi Hua, Deming Dong, Dapeng Liang and Zhiyong Guo*	**141**

Chapter 6	Management of EDTA-Containing Aqueous Effluent: Environmental Concerns and Remedies *Zinnat A. Begum, Ismail M. M. Rahman and Hiroshi Hasegawa*	**163**
Index		**179**

PREFACE

In this book, the authors present current research in the study of the synthesis, uses and environmental concerns of EDTA. Topics discussed include the symbiotic and synergic effects in amide and ester derivatives of EDTA; photo- or radiation-induced degradation of EDTA and its complexes of heavy metals; the use of EDTA root conditioning in regenerative procedures of periodontal tissues; effects of EDTA on adsorption behavior of cadmium onto natural freshwater surface coatings; and the environmental concerns and remedies in the management of EDTA-containing aqueous effluent.

Chapter 1 - The reaction of the dianhydride of ethylenediaminetetraacetic acid (EDTA) with appropriate reagents (mainly amines and alcohols) is an easy synthetic route to obtain a variety of (from mono- to tetramer-) derivatives. Bis(imides) are also common derivatives. EDTA cannot be seen as a simple bridge or tether between the substituents as symbiosis and synergic phenomena are present in the new derivatives. EDTA increases the applications of the linked moieties by transferring its own characteristics, enhancing, for instance, the solubility of hydrophobic moieties or providing the chelating ability. Similarly, the substituents (among others: long alkyl chains, proteins, polymers, or moieties with applications in Supramolecular Chemistry) afford for applications of EDTA in new fields. In the new compounds, synergic cooperative effects, caused by the co-existence at the same entity of the moieties, may also be an expected consequence. Enzymes mimicry or scissor molecules are nice examples.

Chapter 2 - High dielectric permittivity has been found on the oxide $CaCu_3Ti_4O_{12}$ (CCTO), in the order of 10^5 at 1 kHz and room temperature, being nearly constant to temperatures until 300 °C. These characteristics lead this material to be useful in a high number of technological applications, including the miniaturization. The physical explanation for the high dielectric permittivity behaviour is actually an open question. Some researchers suggest that this behaviour is intrinsic, while others attribute to extrinsic effects. Within this work the authors report the synthesis and characterization of CCTO powders prepared by two different wet-chemical methods: ethylenediaminetetraacetic acid-gel method (EDTA) and modified Pechini (MP) process. In both cases, almost single-phase CCTO fine powders were obtained from the calcination of the gel precursors at lower temperatures and shorter times than those reported for the conventional solid state reaction process. Also thin films were produced using spin-coating technique, depositing the obtained gels on Si(111) substrates. The structure of the calcinated powders, sintered pellets and thin films were characterized by X-ray diffraction and scanning electron microscopy. Impedance spectroscopy measurements were performed on the sintered pellets and films, in the frequency range from 40 Hz to 110 MHz, at room temperature. The results are discussed and compared. High frequency relaxation data was fitted with Cole–Cole model and the calculated relaxation parameters were correlated with morphological and structural properties.

Chapter 3 - EDTA is present in a variety of waste, of which one of the most dangerous is liquid radioactive waste produced for years in nuclear power generation, defense applications, and various research institutions or medical applications of radionuclides. Some of these wastes (e.g. decontamination solutions or evaporator concentrates) contain, in addition to the radionuclides, also strong organic complexants like EDTA, generally used as decontamination and washing solutions. The presence of the organic complexants can cause gas generation, microbiological activity, void formation, elevated leachability and simultaneously higher mobility of cationic contaminants and thus can negatively influence the quality of the final forms of treated waste.

For the treatment of organic-bearing aqueous waste, photocatalytic degradation by 254 nm UV radiation with a TiO_2 photocatalyst can be used. Systematic study of the decomposition of EDTA and its metal complexes has been carried out. It was found that while pure EDTA can be degraded with high efficiency, the decomposition of complexes is much more complicated. Separation of the radionuclides by ion exchange can follow the photocatalytic degradation of organic complexes and complexants in a multistage process developed and tested for the treatment of this type of wastes.

Also the utilization of ionizing radiation (by accelerated electrons or γ-irradiation) could lead to degradation of organic complexants and consequently to minimization of the volume and solubility of waste containing metal complexes with strong organic agents. E.g. EDTA and citric acid are radiolytically destroyed by •OH radicals with high rate constants achieving values about 10^9 $dm^3.mol^{-1}.s^{-1}$. Therefore the complexing agent could serve as an •OH radical scavenger enabling the radiation reduction of metal ions released after the destruction of complexes. Another possible mechanism is the precipitate formation consisting of products of organic complexant degradation and metal.

Chapter 4 - Periodontitis is a chronic destructive inflammatory disease caused by bacteria present in dental plaque. This condition induces the breakdown of tooth supporting apparatus, and the ultimate result of the untreated disease is tooth loss. The primary goal of periodontal therapy consists not only to arrest the tissue destruction caused by periodontal disease but also to regenerate the tooth supporting tissues, including alveolar bone, periodontal ligament and cementum. Several procedures have been proposed to enhance regeneration of periodontal tissues, including chemical root surface conditioning. The concept of root conditioning has been based on the perceived need for decontamination of the root surface devoid of attachment apparatus and for exposure of collagen fibers. Periodontitis-affected root surfaces are hypermineralized and contaminated with bacteria and inflammation-inducing bacterial components, such as endotoxin. In addition, the smear layer formed during mechanical root instrumentation has been suggested to act as a physical barrier, inhibiting new connective tissue attachment to the root surface. Therefore, a number of varying pH agents have been used to eliminate the factors that render the root surface bioincompatible for periodontal tissues regeneration. Root conditioning with a neutral pH agent such as EDTA appeared to produce in a short-term perspective a more biocompatible surface compared to conditioning at low pH. Etching at low pH may induce an immediate necrosis while etching at neutral pH preserves the integrity of exposed collagen fibers and the vitality of the adjacent tissues. EDTA root conditioning removes the root surface-associated smear layer, widening dentinal tubule orifices, and exposing collagen matrix when it is applied after scaling and root planing. The aim of this article is to present an overview of the biological effects of EDTA conditioning for root surface modification and to discuss the use of this chemical agent in regenerative procedures of periodontal tissues.

Chapter 5 - Natural surface coatings play important roles in determining behavior of pollutants in aquatic ecosystems. Natural surface coatings usually possess remarkable heavy metal adsorption abilities. EDTA is one of the common artificial ligands encountered in natural waters. It is also a commonly used surrogate for natural organic matters, to assess the roles of natural organic matters in natural aquatic environments. In this study, the effects of EDTA on adsorption behavior of cadmium onto natural freshwater surface coatings were investigated, including adsorption kinetics, adsorption isotherm and pH-edge adsorption. The influences of solution pH and EDTA concentration were also were investigated. The results indicated that EDTA usually has considerable effects on adsorption of cadmium to natural surface coatings. EDTA made the adsorption process approach equilibrium more quickly. EDTA decreased the adsorption of cadmium to surface coatings, especially when the ratio of concentration of EDTA to that of cadmium was high, or when the pH of adsorption solutions was high. The dissolution of some surface coating components was also observed. These effects of EDTA could be explained in terms of the competition between EDTA and the surface coatings for available cadmium, the converting of positively charged cadmium ions to neutrally or even negatively charged species, and the dissolution of manganese oxides from the surface coatings.

Chapter 6 - Ethylenediaminetetraacetic acid, commonly abbreviated as EDTA, has received widespread acclaim among the aminopolycarboxylate chelants for its superior metal binding capability to generate soluble metal-EDTA complexes having diminished reactivity. EDTA has been and continues to be used in a variety of industrial applications, or otherwise. The objective is either to mask the hindering effect of unwanted metal-species or the separation of the toxic metals. Management of EDTA-containing aqueous effluents has received considerable attention because it not only possesses the potential to affect metal bioavailability but also can remobilize the metals from sediments and aquifers. Besides, the persistent nature of EDTA in the aqueous environment due to the slower rate of biodegradability is the largest issue of concern considering the consequent risk to the potable water resources. The proposed remedies are primarily focused on the destruction of EDTA in the aqueous effluent before being discharged to the natural surroundings. The recovery of EDTA from the effluent for subsequent application is also proposed as a remedial option to minimize the probable risk from EDTA. Within the scope of this chapter, the authors provide a brief overview, in light of the recent findings, about the possible environmental concerns related to the discharge of EDTA-containing effluent and the proposed options to compromise the risks.

In: EDTA
Editor: Andris Molnar

ISBN: 978-1-62808-146-6

Chapter 1

SYMBIOTIC AND SYNERGIC EFFECTS IN AMIDE AND ESTER DERIVATIVES OF EDTA

Francisco Meijide, Juan V. Trillo, Santiago de Frutos, Aida Jover, José Vázquez Tato[1*], Victor H. Soto[2] and Luciano Galantini[3]

[1]Universidad de Santiago de Compostela, Facultad de Ciencias, Departamento de Química Física, Lugo, Spain
[2]Universidad de Costa Rica, Escuela de Química, Centro de Investigación en Electroquímica y Energía Química (CELEQ), San José, Costa Rica
[3]Università di Roma "Sapienza", Dipartimento di Chimica, Roma, Italy

ABSTRACT

The reaction of the dianhydride of ethylenediaminetetraacetic acid (EDTA) with appropriate reagents (mainly amines and alcohols) is an easy synthetic route to obtain a variety of (from mono- to tetramer-) derivatives. Bis(imides) are also common derivatives. EDTA cannot be seen as a simple bridge or tether between the substituents as symbiosis and synergic phenomena are present in the new derivatives. EDTA increases the applications of the linked moieties by transferring its own characteristics, enhancing, for instance, the solubility of hydrophobic moieties or providing the chelating ability. Similarly, the substituents

* Corresponding author: jose.vazquez@usc.es.

(among others: long alkyl chains, proteins, polymers, or moieties with applications in Supramolecular Chemistry) afford for applications of EDTA in new fields. In the new compounds, synergic cooperative effects, caused by the co-existence at the same entity of the moieties, may also be an expected consequence. Enzymes mimicry or scissor molecules are nice examples.

INTRODUCTION

During the fifties of previous century, the number of published papers with the "EDTA" concept increases steadily and quickly. By 1950 only three papers are referenced by "Scifinder", while by 1960 this number was higher than 600. Among other reasons, chelones (in which EDTA was the chelating agent) were used thoroughly for the analysis of metals [1]. From that time, many other applications have been found for EDTA and derivatives as, for instance, chelation therapy [2]. The utility of low molecular weight gadolinium complexes as contrast enhancing agents for magnetic resonance imaging (MRI) is another example [3].

A simple way of incrementing the high number of applications of EDTA is to link new residues which transfer their inherent properties into the chelating agent. This linking can easily be performed thanks to the four carboxylic groups in the EDTA structure. If appropriate substitutions are made, the chelating properties of EDTA are essentially preserved and completely new applications are transferred to the new compound. For instance, during the seventies long hydrophobic alkyl chains were bonded to EDTA through amide or ester bonds, and a new type of (chelating) surfactants was born. Later, other moieties, as cyclodextrins (which can form inclusion compounds with many organic molecules) were also linked to EDTA incorporating this chelating residue into Supramolecular Chemistry. These are topics which constitute the scope of this chapter. The chelating ability of unmodified EDTA [4] is not reviewed in this chapter, neither amides or ester with short alkyl chains or aromatic rings. References [5-10] are examples of uncovered papers.

For the synthesis of the new compounds, EDTA **1** is usually converted into its dianhydride **2** which is frequently used as the starting material for further reactions. Several recipes may be found in the literature cited below. Partial hydrolysis of **2** forms the monoanhydride **3**.

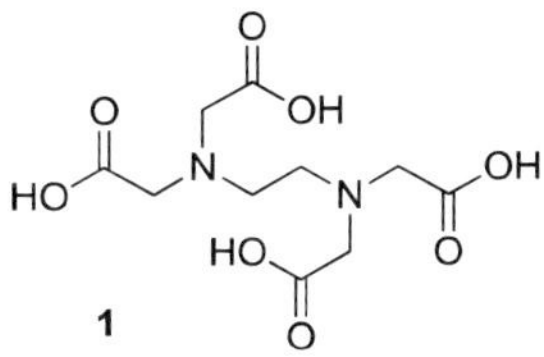

1

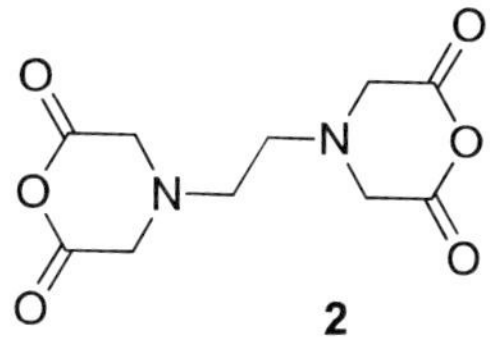

2

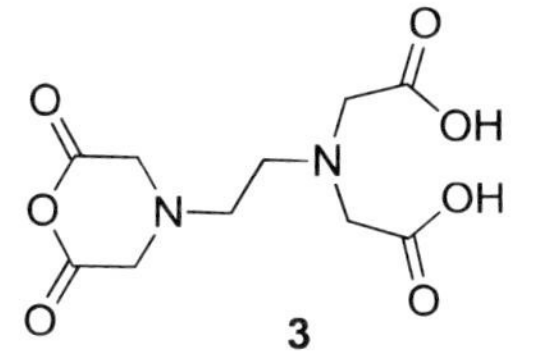

3

Typically, compound **3** is used to obtain monoamides **4** (with three free carboxylic groups) by reacting with amines, while **2** is used to obtain symmetric bis(amides) **5** (with two free carboxylic groups), bis(imides) **6** (without free carboxylic groups) and symmetric (R=R') or asymmetric (R≠R') tetraamides **7**.Triamides are also known [11]. Compounds with free carboxylic groups can be protonated in different degree allowing to obtain compounds with different charges. This can also be obtained by complexing ions with different charge. Both phenomena can be done simultaneously.

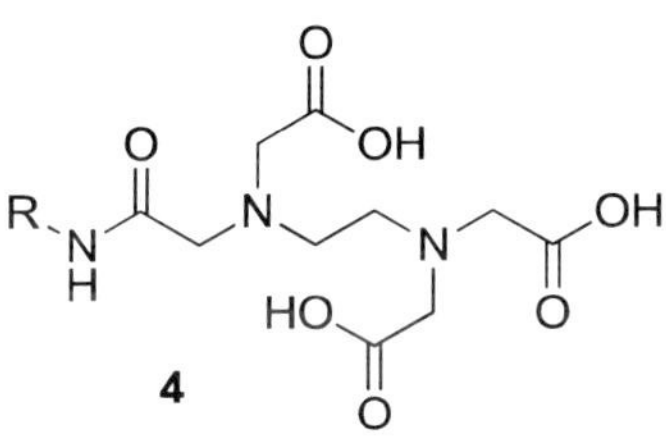

4

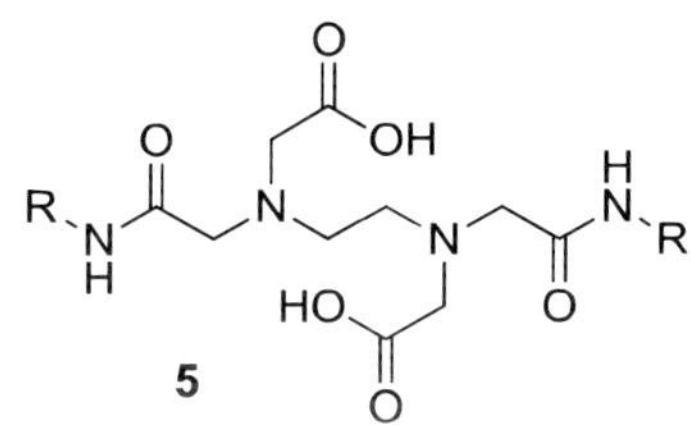

5

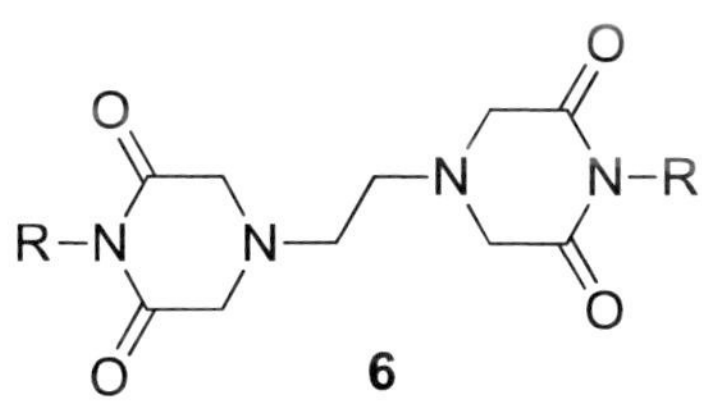

6

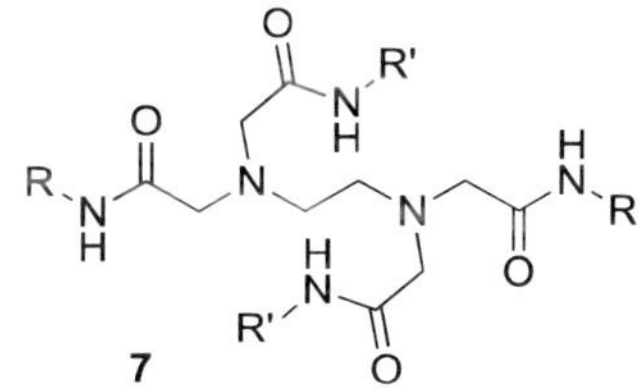

7

Analogously, the reaction of **3** or **2** with alcohols yields monoesters **8** and diesters **9**, respectively. Triesters, monoamide triesters [12], and tetraesters **10** have also been reported. Tetraesters have been used as precursors of tetraamides [13] and to complex some metal ions [14]. As previously, not all of them will be reviewed here, and, with a few exceptions (the artificial phospholipid design by Furuta et al. [15] is an example), neither will be considered those derivatives with substituent groups in the core ethylene moiety.

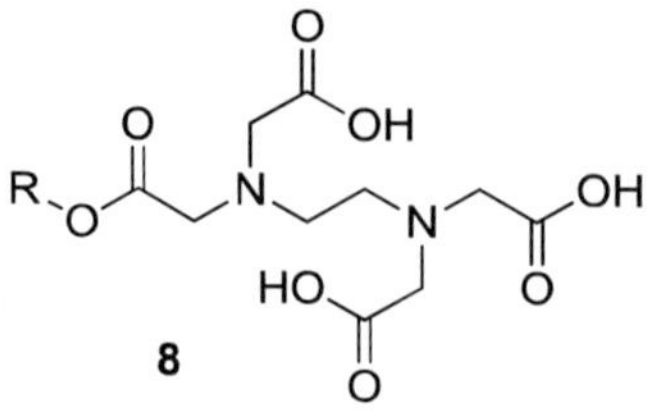

8

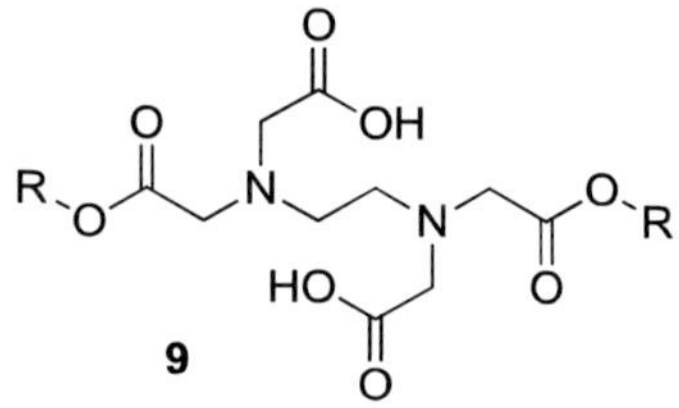

9

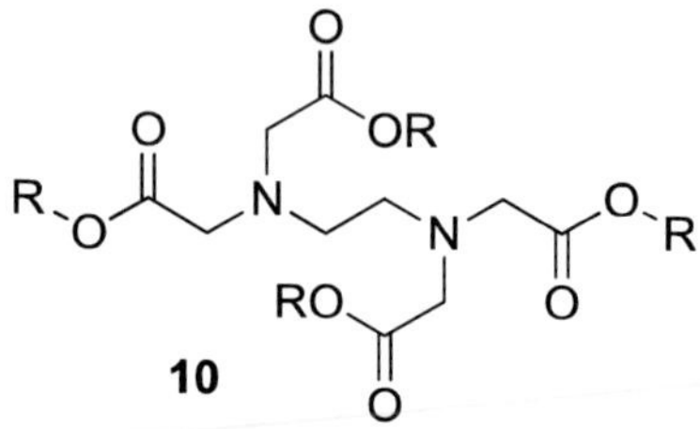

10

Compounds are reviewed according to previous indicated substitutions, although exceptions are made for substituents with specific structures or properties (bile acids, proteins...) which are compiled independently.

MONOAMIDE AND MONOESTER DERIVATIVES (4 AND 8)

Typically, in compound **4**, R is an alkyl chain with *n* carbon atoms. If the chain is saturated R obeys the formula C_nH_{2n+1} (**4.1**) Similar formulae may be written for unsaturated chains with different degrees of saturation.

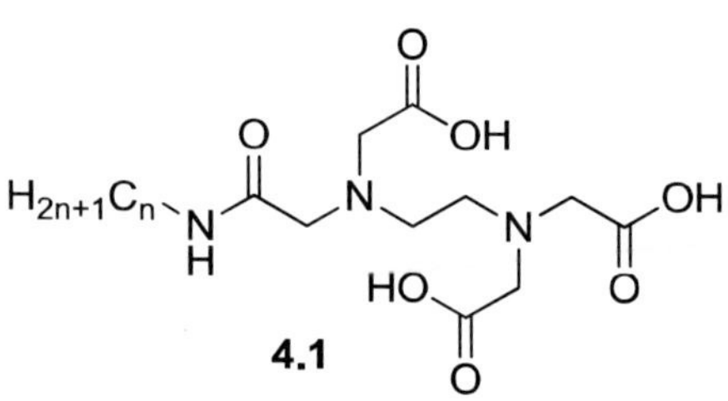

4.1

Monoalkylamides **4.1** can be prepared in good yields from an alkyl amine and monoanhydride **3** according to the following general synthetic scheme [16]

$$C_nH_{2n+1}\text{-}NH_2 \quad + \ \mathbf{3} \longrightarrow \mathbf{4.1}$$

An example of previous monoalkyl amides corresponds to the dodecyl group, i. e., R= $C_{12}H_{25}$, obtained, among others, by Jaeger et al. [17]. Many other compounds with *n* in the range 2-40 have been synthesized as Table 1 shows. In this Table, monoalkyl ester chelators **8.1** [18] and alkyl amines **11** [19], which are closely related to previous compounds, are included as their properties and uses are similar to those of the monoalkyl amide derivatives.

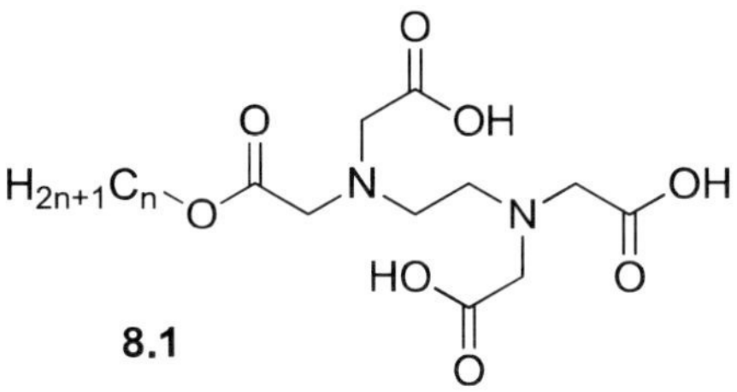

8.1

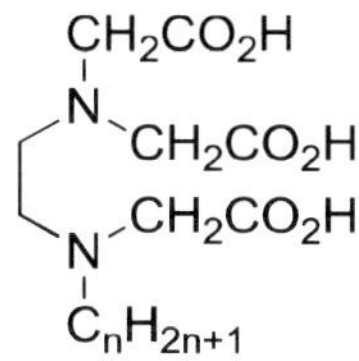

11

Table 1. Examples of monoalkyl amides, monoalkyl esters, and some closely related compounds derived from EDTA

R = C_nH_{2n+1}	**Reference**	**Related structures**	**Reference**
n = 10, 12	(*16*)	**8.1** *n*=10, 12, 14, 16, 18 or oleyl alcohol.	(*20*)
n = 12-18	(*21*)	**8.1**, *n*=8, 10, 12	(*19*)
n=6-22	(*22*)	**8.1** *n*=4,10	(*18*)
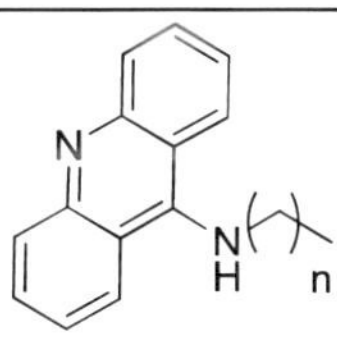 n=2-12	(*23-24*)	**11** *n*= 8, 10,12	(*19*)
n=2-40	(*25*)		
n=12	(*17*)		
n=8, 18	(*26-27*)		
n=8,10,12	(*19*)		
n = 18	(*28*)		
$(CH_2)_{12}OH$	(*19*)		
$(CH_2)_{12}N^+Me_3$, Br^-	(*19*)		

Takeshita et al. [16] prepared monoalkyl amides **4.1** (*n* = 10, 12, 14, 16 and 18) with yields above 50%. This contrasts with the equivalent reaction between EDTA dianhydride and alcohols which gave EDTA monoalkyl esters **8** in poor yield (~13%) due to dialkyl esterification [20]. These amide derivatives form 1:1 chelates with Al, Fe, Co, Ni, Cu, and Zn. The chelates and the sodium salts of the chelating agents have good surfactant properties, especially dispersing power and emulsifying power. For instance, the chelates **8.1** (*n*=4, 10) give values of 30.1-31.4 mN/m for the surface tension in aqueous solutions. The interfacial tension is 0-5.3 mN/m against kerosine, and sodium salts of the chelates have good dispersing power for TiO_2. The monoesters, prepared from R-OH alcohols (R = C_nH_{2n+1}, *n* = 10, 12, 14, 16, 18, and oleyl) were used with nitrates of Al, Ca, Fe, Co, Ni, Cu, Zn, Cd, and Pb to prepare chelates which good surface activities. The Al-chelate of the monooctadecyl ester is the best dispersant for TiO_2 and is among the best emulsifiers for kerosine.

Compounds **4.1** (*n* = 8, 12) have been investigated by Novak et al. [29]. Table 2 resumes the pK_a values for the acidity constants (sodium salt derivatives) as well as the stability constants, pK_{MA}, for the Ca(II) and Mg(II) complexes. The results indicate that the introduction of the hydrophobic alkyl chain results in significant changes in those parameters. pK_a values for the dissociation of the tertiary amino groups of bis(alkyl amides) have been measured by Danil de Namor and Pacheco Tanaka [30] and Platas-Iglesias et al. [7].

Novak et al. carried out several biological tests and detected that these compounds cause irritation of the stomach wall, have some irritation effects on skin, inhibit the growth of several microorganisms, and are nonmutagenic. These compounds are good surfactants, have medium foaming ability, high washing efficiency, and a wettability which increases with the concentration of the compound. Novak et al. compiled formulations of washing means, showing comparable washing ability to commercial available powder detergents, and a very low value of incrustation of the fabric. These properties were related to the chelating effect of these EDTA alkyl amide derivatives. The set of compounds **4.1** (*n*=12-18) have been patented as components in the formulation of detergent compounds for laundering (particularly cotton fabrics) at 30-60ºC [21], as well as foaming cleaners (*n*=6-22) [22]. Other related patents have been published by Prochazka et al. [31-32]. Similarly, He et al. [25] have patented compounds **4.1** (where *n* is 2 to 40 as well as unsaturated chains with *n*=3-40) for liquid personal washing compositions.

Sodium salts of **4.1** (*n*=8, 18) have been proposed as stabilizers of enzyme components in liquid and paste laundry detergents [26-27].

The Co(III) complex **12** was obtained by the reaction [17]

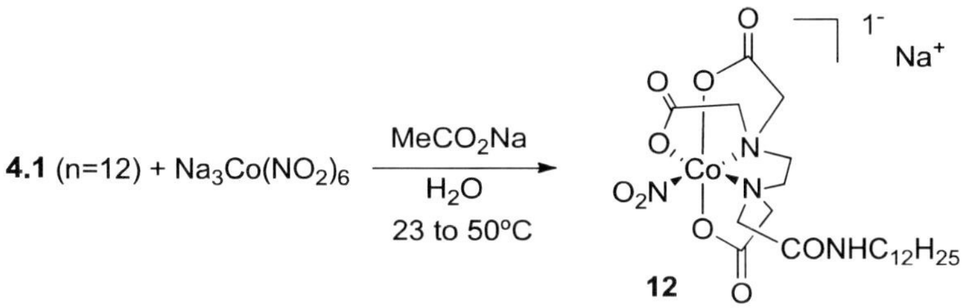

Five of the six coordination sites of octahedral Co(III) are occupied by ligand **4.1** (*n*=12), and the sixth is occupied by a nitro ligand. Both **12** and **4.1** (n=12) (in form of trisodium salt) behave as surfactants with Krafft temperatures of 54 and 23 °C, respectively. The critical aggregation concentration, *cac* (*cmc* in the case of micelles), in H_2O (at 23 °C) of **12** and the trisodium salt of **4.1** (*n*=12), were determined by surface tension measurements. Values of $(5.3\pm0.3)\times10^{-4}$ and $(4.4\pm0.2)\times10^{-4}$ mol dm^{-3}, were obtained. It is to say, the effect of the complex head on *cac* is not very strong. However, the effect on size and geometry of aggregates is much stronger since surfactant **12** aggregates in H_2O at 50ºC to form giant vesicles and rod/tubules (as determined by phase-contrast optical and cryo-high-resolution scanning electron microscopy) while aggregates of the trisodium salt of **4.1** (*n*=12) in H_2O (determined by dynamic laser light scattering) have a hydrodynamic diameter of 3.5±0.5 nm, a value consistent with the formation of small spherical micelles.

Table 2. pK_a values for the acidity constants for 4.1 (*n* = 8, 12) and pK_{MA} (stability constant) for the chelates with Ca(II) and Mg(II). Values are from Novak et al. [29]

Compound	pK_{a1}	pK_{a2}	pK_{a3}	pK_{a4}	pK_{MA} Ca	pK_{MA} Mg
EDTA	2.00	2.76	6.61	10.26	10.96	8.69
4.1 R = $C_{12}H_{25}$	2.7	4.5	9.5		8.51	6.26
4.1 R = C_8H_{17}	2.77	4.48	9.73		9.82	7.81

New surfactant Co(III) chelates (ligands **4.1** (n=8,10); **11** (n=8,10,12); **4.1** R=$(CH_2)_{12}OH$; **4.1** R=$(CH_2)_{12}N^+Me_3Br^-$) with structures very close to the one of **12** were obtained by Jaeger et al. [19]. The pentadentate ligands correspond to the trianions of EDTA derivatives indicated in the structures, in addition to a nitro ligand.

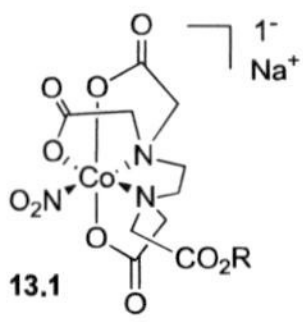
13.1

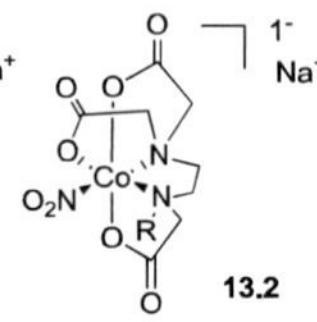
13.2

R: C_nH_{2n+1} n = 8, 10, 12

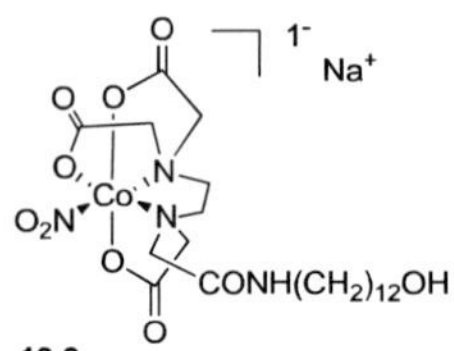
13.3

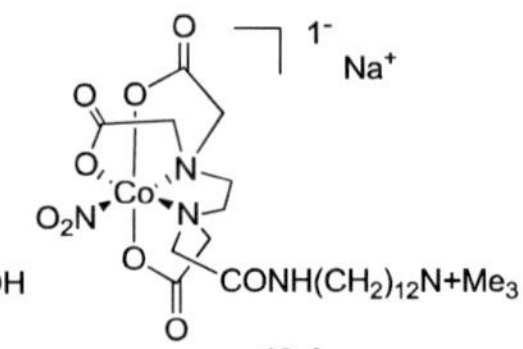
13.4

Compared to surfactant chelate **12**, the new complexes contain long chain ester, alkyl, ω-hydroxy and ω-trimethylammonium groups. The chelate with the ligand R=$(CH_2)_{12}N^+Me_3Br^-$, may be considered a zwitterionic bolaform surfactant, since it contains a negatively charged Co(III) complex at one end of the hydrocarbon chain and a positively charged ammonium group at the other end. The structures of these surfactants were established by ^{1}H and ^{13}C NMR, IR, and UV-vis spectroscopy and their Krafft temperatures and *cac* in water were also measured. The measured *cac* values are from $(5.3\pm0.3)\times10^{-3}$ mol dm^{-3} for **12** and $(3.7\pm0.1)\times10^{-2}$ mol dm^{-3} for **13.2** (n=8). From the slightly broadened signals of ^{1}H NMR spectra of the chelate surfactants, Jaeger et al. have concluded that the surfactants form small aggregates as micelles or small vesicles, in marked contrast to the formation of vesicles and rods by surfactant Co(III) chelate **12**. However the proposed aggregates are far from being fully characterized.

Some compounds with the structure **4.2** (and their esters) have been proposed as antiviral compounds [23-24]. They were synthesized by the reaction of the corresponding amines (carrying the acridinyl residue) with EDTA triethylester. The triester derivative was prepared from the corresponding tetraethyl ester by selective enzymatic monodeesterification using pig liver esterase using the mixed-anhydride method (isobutyl chloroformate/N-methyl-morpholine N-oxide). The saponification of the esters was performed by treatment with aqueous sodium hydroxide. The target compounds were treated with a 2-5 molar excess of Fe(III), lyophilized and incubated with the phages in Tris-buffer in the presence of sodium ascorbate. For biological screening the well established system containing MS2 bacteriophages was used.

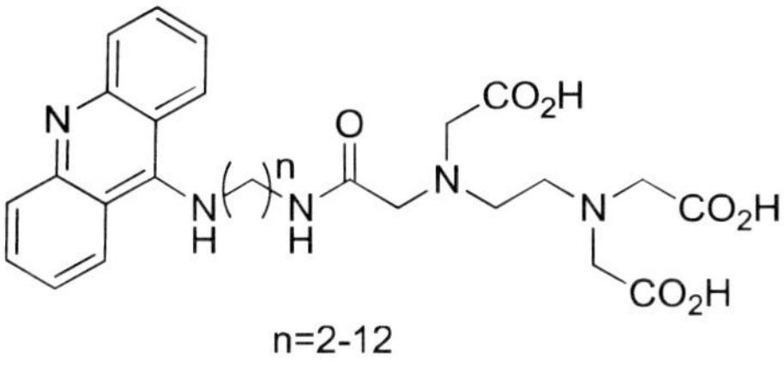

4.2

The length of the spacer exhibits significant effects on the reduction of the phage-titer, the best results being obtained for the compound with n=6. For instance, it is around 10^5 times more effective than the compound with n=2. The inactivation of this virus also depends on the concentration, temperature, time of incubation and the concentration of ascorbate. These compounds do not show any activity in the absence of ascorbate and no significant antipathogenic activity was found upon incubating the MS2 phages with acridine and Fe(III)-loaded EDTA in the presence of ascorbate.

The octadecyl derivative **4.1** (n=18) [28] was labeled with ^{99m}Tc with high efficiency. The organ distribution of the complex was studied in white rats. After 15 min the relative activity in kidney account for 48% of the total amount, reaching 82.5% after 4 hours. When the compound was bound to liposome, the biodistribution 60 min after administration indicates a decrease (approx. 50%) in the relative activity in kidney (compared to the original complex) accompanied by an important increase activity in stomach. Other ^{99m}Tc complexes with EDTA derivatives have been investigated as kidney imaging agents by Havranek et al. [33]. Eckelman et al. [34] affirm that diesters **9.1** and **9.2** strongly chelate radioisotopes such as ^{57}Co, ^{99m}Tc, and ^{113m}In. A broad overview of the current status and future prospects of ^{99m}Tc radiopharmaceuticals has been published by the International Atomic Energy Agency [35].

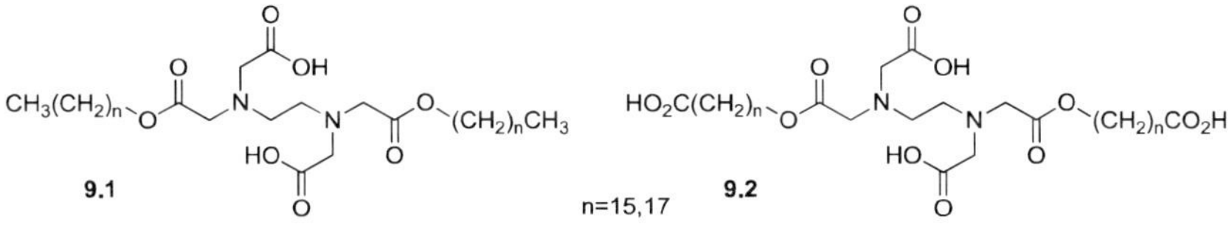

Soil washing is one of the few permanent treatment alternatives for removing metal contaminants, and, for this purpose, EDTA derivatives and their salts are very effective. To overcome some inconvenients of the EDTA application (polarity of the tetraanion, limited solubility of metal complexes in the organic solvents, enhanced mobility of heavy metal, non-selective chelating capacity, etc) [18, 36], some EDTA derivatives have been proposed, even as potential effective agents for the in vivo aluminum removal [37]. Zhang et al. [18] have prepared the two EDTA monoalkyl ester chelators **15** from **14** derivative according to the scheme:

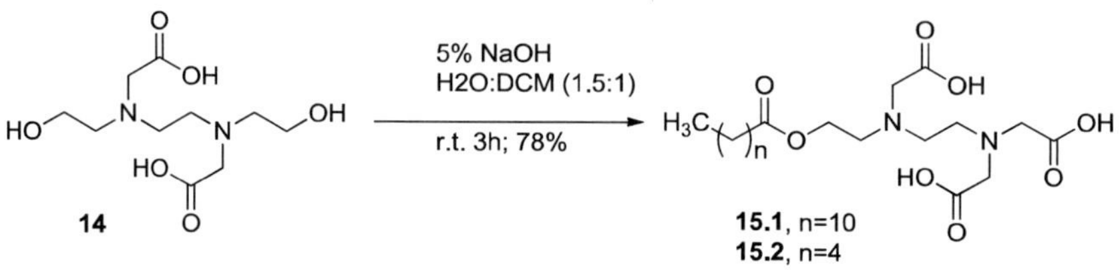

15.1 shows a typical surfactant behavior when surface tension is plotted against the logarithm of concentration, *cmc* being 1.30×10^{-3} mol dm^{-3} at 25 ºC and pH 7. This is not the case of **15.2.** The chelating ability (Ca(II) chelation) decreased with the increase in the alkyl chain length (EDTA > **15.2** > **15.1**), a fact which was attributed to hydrophobic interactions of the alkyl chain. However, above *cmc*, **15.1** achieves a similar ability to **15.2**, indicating that the formation of micelles by **15.1** favors the chelation of Ca(II). No similar significant increase in this ability was observed for **15.2** when its concentration was similarly increased, supporting that it cannot form micelles.

These compounds were studied as potential extractants for oxyanion Cr(VI) and cationic Cu(II). Experiments have shown that **15.1** and **15.2** can extract both ions from humic acid simultaneously. The Cr(VI) extraction percentages by EDTA and **14** were negligible, while those of **15** increased with increasing concentration of the chelator, although **15.1** was more effective. The ability of **15.2** for the extraction of Cr(VI) was mainly attributed to a competitive adsorption mechanism via ion exchange. This would be also the case for 15.1 below *cmc*, but above it, because of its surface active properties that enhance the solubility of humic acid, a mechanism of dissolution of humic acid becomes dominant for Cr(VI) extraction, resulting in a better extraction ability of **15.1**.

The release of Cu(II) from humic acid increased rapidly with increasing concentration of all chelators, being more rapid for EDTA. As for Cr(VI) extraction, the extraction by **15.1** was attributed to both its negative charge and its surfactant properties.

Phospholipids may be used as the carriers of the amino group to react with **2** or **3**. The reaction between **2** and dipalmitoylphosphatidylethanolamine is an example, leading to the formation of the monoamide **4.3** (30% yield) [38]. The compound is insoluble in buffered solutions Tris(0.1 mol dm^{-3})/NaCl (1.5 mol dm^{-3}) and pH = 7.4, but is soluble in NH_4OH/NH_4Cl (0.1 mol dm^{-3}; pH =9.5) buffer. Then it becomes possible to prepare complexes with In(III) or Gd(III) (40% yield), soluble in the mentioned Tris-buffer. Analogously, the derivative of diethylenetriaminepentaacetic acid (DTPA) **16** was also prepared.

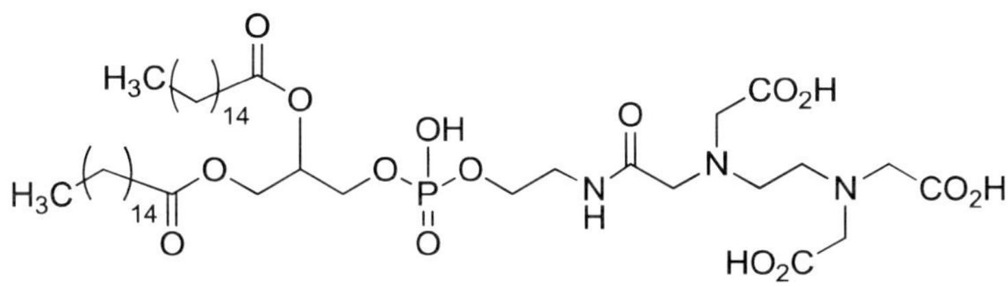

4.3

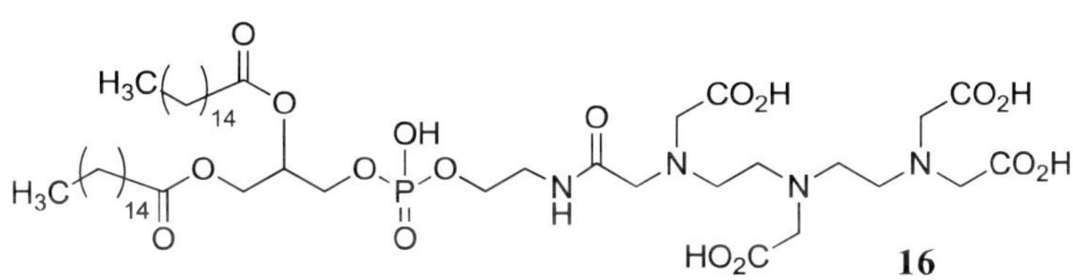

16

In a series of papers, Takeshita et al. [39] have studied the effects of additives, water, and coal types on the coal-oil mixtures (COM) stability. Suitable surfactants were sodium dodecylbenzenesulfonate, disodium salt of **4.1** (n=16), and compound **17**.

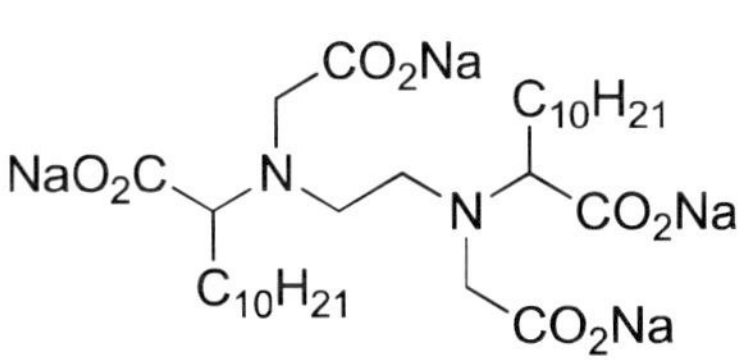

17

COM prepared from other six EDTA derivatives [disodium salts of EDTA monoalkyl esters (alkyl = dodecyl, tetradecyl, and oleyl) and disodium salts of

EDTA monoalkyl amides (alkyl = dodecyl, hexadecyl, and octadecyl)], were much more stable than those prepared from commercial surfactants [40].

Furuta et al. [15] have designed and synthesized the Fe(III) chelating phospholipid **18** as a "scissor" molecule for integral membrane proteins and preliminary investigation of its cleavage activity toward influenza virus membrane protein, hemagglutinin. In spite of no detection of peptide fragments cleaved, the results indicated scission of hemagglutinin which, in contrast, resists cleavage with the simple Fe(III)–EDTA complex. Cleavage of bovine fetuin (a water-soluble glycoprotein) by **18** was also observed although, in this case, the Fe(III)–EDTA complex is effective as well. Other C1 carbon atom derivatives of EDTA can be used to align proteins in solution without the need to add liquid crystalline media [10, 41].

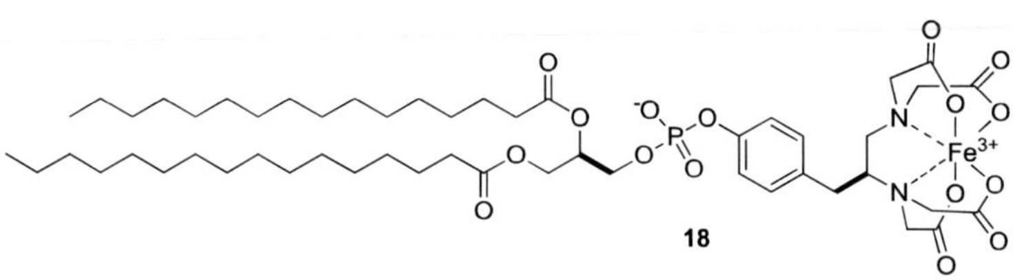

18

BIS(AMIDE) AND DIESTER DERIVATIVES (5 AND 9)

As indicated, bis(amide) derivatives can be obtained by the reaction of **2** with amines, according the following scheme

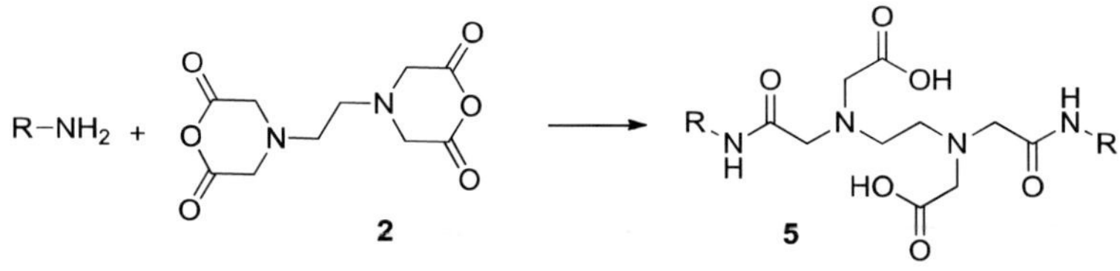

Diester derivatives can be obtained analogously.

Bis(alkyl amides) **5.1** (n=3-18) were prepared by Takeshita et al. [20] as well as bis(dialkyl amides) with the same substituent groups. Takemitsu et al. [42] prepared bis(alkyl amides) **5.1** (n=6-12) by hydrolysis of the corresponding bis(alkyl imides). Some metal chelates were obtained by the reaction of these amides with Fe(III), Cu(II), Zn(II), Ni(II), Co(II) and Al(III) nitrates. The chelate formed by EDTA bis(N-dodecyl amide) and Ni(II) had a good dispersing power for TiO_2 pigments in organic solvents. C-alkyl EDTA chelates have been also investigated by Takeshita and Maeda [43]. The chelate formed with iron sulfate has an octahedral structure and it has good surface activity.

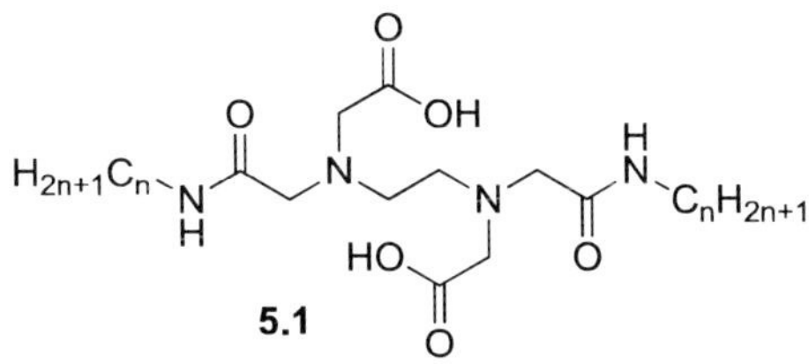

5.1

Secondary amines may be used as well. Yamashita et al. [44] have obtained chelating agents 5 with R_2NH amines where $R=C_nH_{2n+1}$ and n=3-6. The chelates prepared (yields above 90%) with propyl and dibutyl derivatives, as well as the aluminum, cobalt, nickel, and zinc chelates with the dipentyl derivative are soluble in water but all the other chelates are insoluble. Surface and interfacial tensions of the aqueous solutions of the chelates decreased with an increase in the alkyl chain length of the chelates. The sodium salts of the dibutyl, dipentyl, and dihexyl derivatives showed good surface activities. For instance, the surface tensions of sodium salts (10^{-8} mol dm^{-3} at room temperature) in aqueous solution of chelates are 62.6 (dipropyl), 47.7 (dibutyl), 34.7 (dipentyl), and 26.1 mN/m (dihexyl). Dispersing power of the chelates for TiO_2/toluene-butanol (1:1) system increased with a decrease in the alkyl chain length. The subject was reviewed in 1990 by Maeda and Takeshita [45].

Goto et al. [46-47] have obtained **5** ($R=C_{18}H_{35}$, oleyl) derivatives and used them to enhance the mutual selectivity of rare earth metals in liquid-liquid extraction. These agents have a surfactant-like interfacial activity and are strongly adsorbed at the oil-water interface. For the studies, aqueous solutions of three rare earth metals (Er, Y and Yb) were prepared. The organic solvent was toluene and 2-ethylhexylphosphonic acid mono-2-ethylhexyl ester **19** as extractant. The addition of the complexing agent had no influence on the extraction equilibrium constant (the values being 110, 8.4 and 5.8 for Yb, Er, and Y, respectively). The selectivity of Y over Er was significantly enhanced by the addition of a small amount of **5.1** (n=18) because the extraction rate of Y selectively decreased. The experimental results have also shown a synergic effect caused by the interaction between the metal ion and the complexing agent at the oil-water interface.

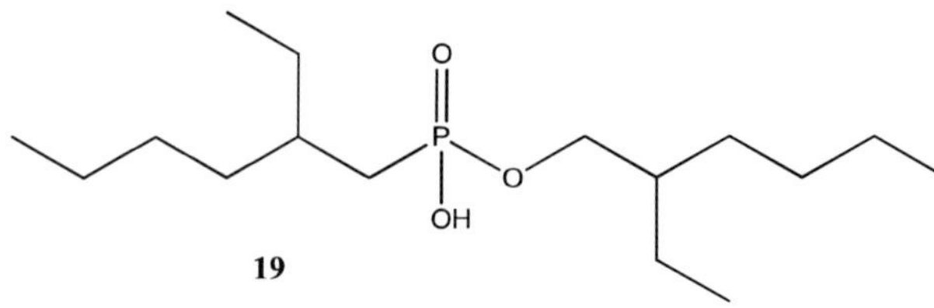

19

BILE ACIDS

Bile acids are at the focus of intensive research. These compounds are interesting because of their biological function, their therapeutic applications, and, in recent years, their extensive use in supramolecular chemistry, materials chemistry and nanotechnology [48-54]. Bile salts exhibit a facial amphiphilicity derived from the existence of a hydrophilic side (α-face, concave lower side) and a hydrophobic side (β-face, convex upper side). The side chain with a carboxylic group or an amide group also contributes to the hydrophilicity of the compound. As a consequence, they are good surfactants and form aggregates in water as long as the concentration is above a critical one [55-56]. Compounds **20** correspond to cholic and deoxycholic acid structures.

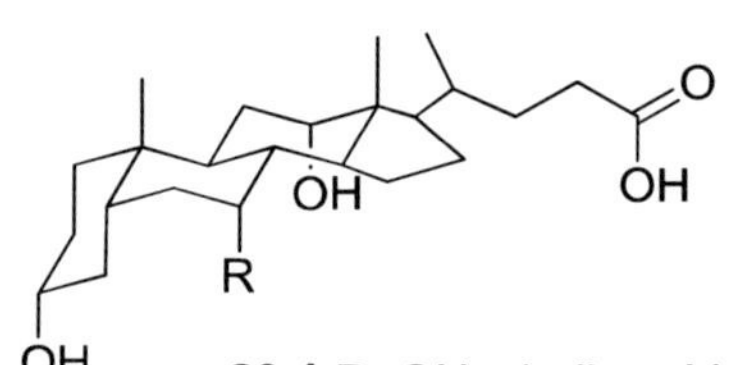

20.1 R=OH, cholic acid
20.2 R=H, deoxycholic acid

There are several examples in which natural bile acids have been linked to EDTA for different purposes. For instance, bile acids are involved in the enterohepatic circulation while metal complexes of polyaminopolycarboxylate chelators are being used as radiopharmaceuticals and as MRI contrast agents. The EDTA-bile acid derivative could be used to deliver a heavy atom (as $^{111}In(III)$ into the hepatobiliary system in animals. With this aim in mind, Betebenner et al. [57] have synthesized the conjugate **21** in which cholic acid is linked to EDTA via an ll-atom spacer. The compound was obtained by reacting the succinimidyl ester of cholic acid with the amine formed by reaction of a benzyl isothiocyanate derivative of EDTA with N-(*tert*-butoxycarbony1)ethylenediamine and subsequent deprotection.

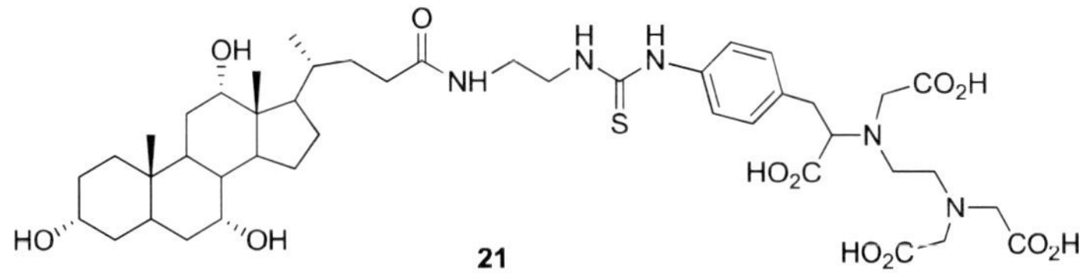

21

From potentiometric titration studies, pK_a values of 9.84, 5.44, and 3.44 were found, which may be compared with those in Table 2. The EDTA moiety of **21** retained the ability to form chelates. Its complexes with La(III), Gd(III), and Cu(II) were too insoluble to permit potentiometric measurements, but it was possible to obtain data for chelates with Zn(II) and Lu(III), having values for the stability constants of log K = 15.9 and 13.9 for $[Lu\text{-}\mathbf{21}]^-$ and $[Zn\text{-}\mathbf{21}]^{2-}$ complexes, respectively. Under similar experimental conditions the stability constants for the 1:1 complexes of EDTA with Lu(III) and Zn(II) are 19.8 and 16.5, respectively. Thus, conjugation reduced these values by ca. 3-4log K units. The stability constant for the $[In(III)\text{-}\mathbf{21}]^-$ was not determined but the one for $[In(III)\text{-}EDTA]^-$ has been reported to be log K=25.8±0.2 at 20 °C [58], a fact which suggests that log K value for $[In(III)\text{-}\mathbf{21}]^-$ is >20.

The in vivo studies into normal mice do not evidence any chelate instability, show rapid clearance of the chelate from the blood, and intense early localization in the liver, with subsequent excretion from the liver into the gastrointestinal tract being complete within 1 hour after injection. Results indicate that the cholic acid moiety is essential for efficient hepatic uptake, i. e., the conjugation of EDTA to cholic acid allows a controlled delivery of EDTA chelates to parenchymal liver cells that otherwise exhibit predominantly extracellular distribution and renal clearance.

The hydrophobic/hydrophilic balance of bile acids [59] can be modified by attaching appropriate groups in their structures. For instance, attachment of hydrophobic groups as adamantyl or *p-tert*-butylphenyl to cholic acid (compounds **22** and **23**) reduce *cac* of cholic acid [60] by 30 and 50 times, respectively (compounds in their carboxylate forms).

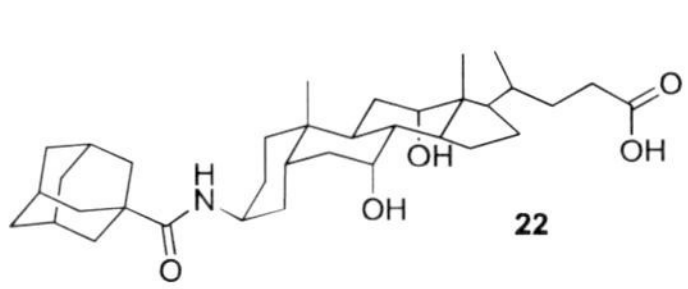

22

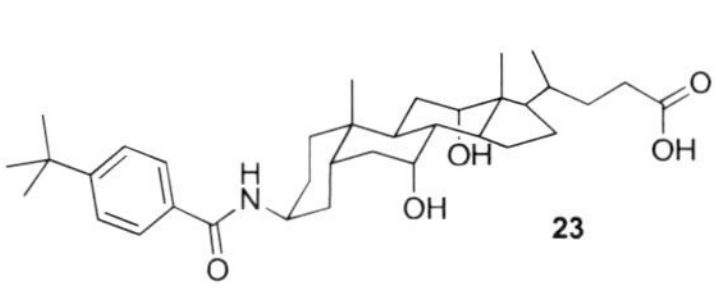

23

By the contrary, the attachment of hydrophilic groups increases *cac*. For instance, the *cac* value for **24** [61-63] is 0.0699 mol dm^{-3} [64] which is five times higher than that reported for sodium cholate [60, 65].

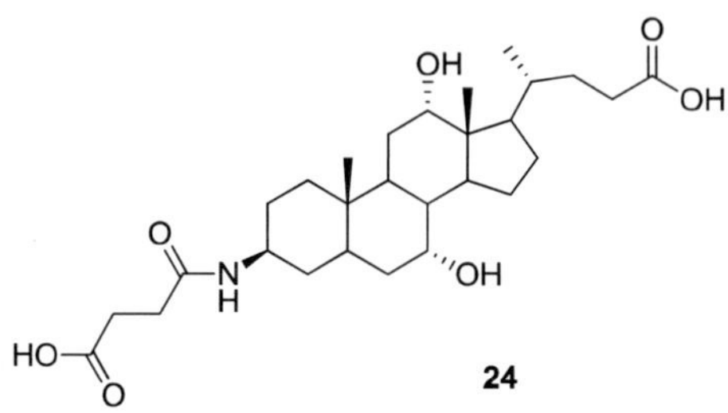

24

It is instructive to compare the surfactant behavior of the two EDTA derivatives **5.2** [66] and **5.3** [67]. Both compounds have two deoxycholic residues linked to EDTA through amide bonds. In **5.2**, the deoxycholic acid was first transformed in its C24 amino derivative [68] and the steroid residues are bonded to the bridge through this C24 carbon atom. Consequently, two carboxylic groups are lost. This compound is the first example of a *gemini* steroid surfactant. In **5.3a**, the deoxycholic acid residues are linked through a nitrogen atom bonded to C3. The C3 amino derivative [69] was synthesized according to Anelli et al. [70]. From surface tension measurements a threshold concentration equal to 4.6×10^{-6} mol dm^{-3} was measured for **5.2**, a value which is 3 orders of magnitude lower than the average *cmc* $(4.4\pm2.9)\times10^{-3}$ mol dm^{-3}, calculated for **20.2** (see reference [56] for compiled values), indicating that the new compound is a much better surfactant than the natural bile salt. For **5.3a** a value of 1.96×10^{-2} mol dm^{-3} was measured for the *cmc*, which is similar to the one for sodium cholate **20.1**. For **5.3b** the measured *cac* in water was 6.7×10^{-3} mol dm^{-3}, a value which is very close to published *cmc* values for sodium dexycholate **20.2** [56, 60].

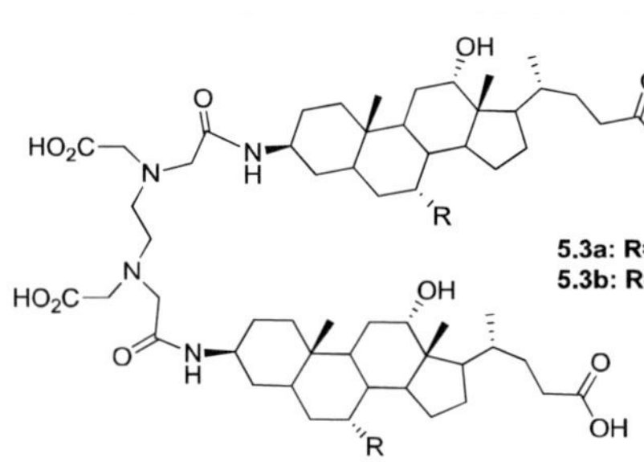

5.3a: R=H
5.3b: R=OH

The observed *cac* values are in the order **24** > **20.1** ≈ **5.3a** for cholic acid derivatives and **20.2** ≈ **5.3b>>5.2** for deoxycholic derivatives. For comparison, the number of hydroxy groups (at C7 and C12 carbon atoms) must be considered. Previous values suggest a rough correlation of the hydrophobic/hydrophilic balance of the surfactant on the n_s/n_c ratio (n_s being the number of steroid groups in the bile acid derivative and n_c the number or negative charges when the acids are in salt form). Therefore it would be expected that at high pH values **21** (n_s/n_c=4) is more hydrophilic than cholic acid **20.1**. Unfortunately, Betebenner et al. [57] do no report the surfactant properties of **21**. However other factors must be also taken into consideration as the comparison between **20.2** and **5.2** illustrates. Since both compounds have the same value for the n_s/n_c ratio, previous results suggest that the stronger hydrophobicity of **5.2** is related to the difference in the surface areas of both compounds if they were lying flat on the surface since **5.2** has two hydrophobic steroid groups and **20.2** only one. Effectively, values of 220 and 90 $Å^2$ have been obtained for **5.2** and **20.2**, respectively**.** Attending to the n_s/n_c ratio, **24** and **5.3a** should have similar *cac* values and higher than the one for **20.1**. However, attending to the number of steroid residues in the compound, the tendency should be **20.1>5.3a**. The observed tendency suggests that the hydrophobic surface effect due to the existence of two cholic residues in **5.3a** is cancelled out by the increase in the number of carboxylate groups respect to **20.1**.

The surfactant behavior of **5.3(a** and **b)** and the ones of their bile salt residues, **20.1** and **20.2,** can be summarized as follows: (i) the molecular areas at the interface for the new surfactants are fairly close to twice the value for a single molecule in a monolayer of natural bile salts; (ii) the environment where pyrene is solubilized is very apolar, as in natural bile salt aggregates; (iii) Gibbs free energies (per steroid residue) for micellization are not far from published values for **20.1** and **20.2**, and the differences can be understood on the basis of the charges in the bridge; and (iv) as for **20.1** and **20.2**, aggregates have rather low aggregation numbers. The fraction of bound counterions in **5.3 (a** and **b)** (*ca* 0.5) is higher than in sodium cholate (close to zero) and sodium deoxycholate (=0.32). Thus the proposed structure for the aggregates of **5.3 (a** and **b)** is based on the disk-like model [71] accepted for the small bile salt aggregates of natural bile salts. However this cannot be the case for **5.2** for which it has been proposed a coexistence of small micelles (average aggregation number =12) and vesicle-like aggregates (aggregation number $=5.5\times10^4$) in a metastable state.

The importance of the hydrophobic surface effect (related to the increase of the number of hydrophobic groups, n_s) and the n_s/n_c ratio can also be found in adamantyl derivatives. For this purpose we will compare the surfactant behavior of adamantyl acetic acid **25** and the adamantyl dimer **5.4** [72]. As previously, the properties are measured with compounds in their acetate forms. Adamantyl derivatives are compounds of interest because of they have a high biological activity including antiviral properties [73-74].

The two compounds **25** and **5.4** have the same n_{Ad}/n_c (Ad: adamantyl) ratio but **5.4** has two hydrophobic adamantyl groups and **25** only one. Because of the higher number of hydrophobic groups it would be expected that **5.4** is a better surfactant than **25** as it has been found experimentally. The surface tensions of **25** and **5.4** at 0.02 mol dm^{-3} are 62 and 49 mN/m, respectively. In comparison to previous derivatives, it would be expected that compound **26** with n_{Ad}/n_c=2, will stronger reduce the surface tension of water. Unpublished work by Trillo [75] corroborates this prediction as a 1.0×10^{-3} mol dm^{-3} solution of **26** has a surface tension of 42.5 mN/m.

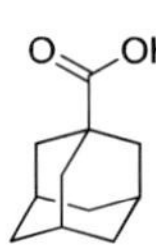

25

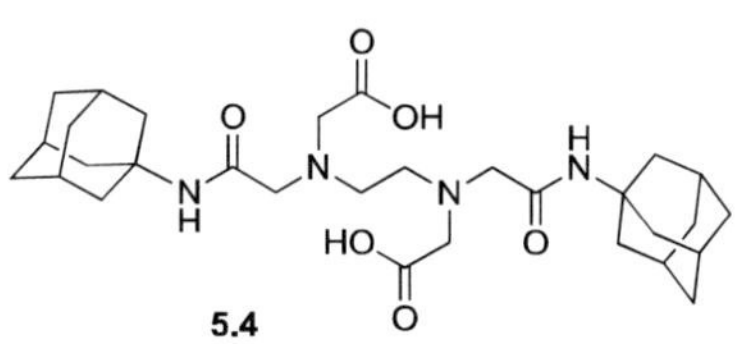

5.4

26

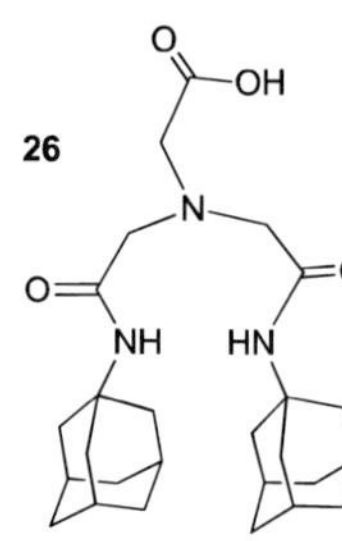

Compound **5.4** illustrates two other useful properties of EDTA when is linked to other moieties. Adamantyl dimers (for example **27** derivatives) are insoluble in water [76] but **5.4** has a high solubility. It is to say, EDTA can be used to enhance the solubility of strong hydrophobic groups. The second property is related to its use in supramolecular chemistry.

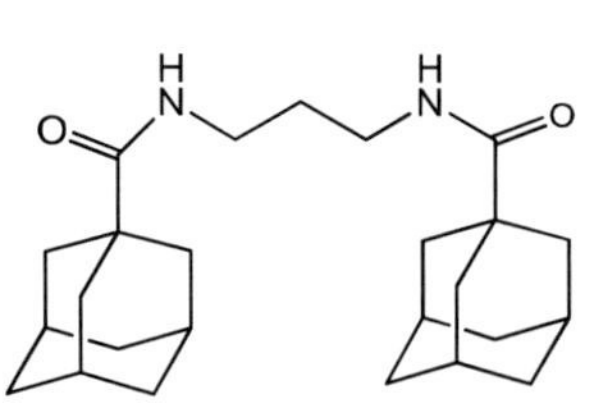

27

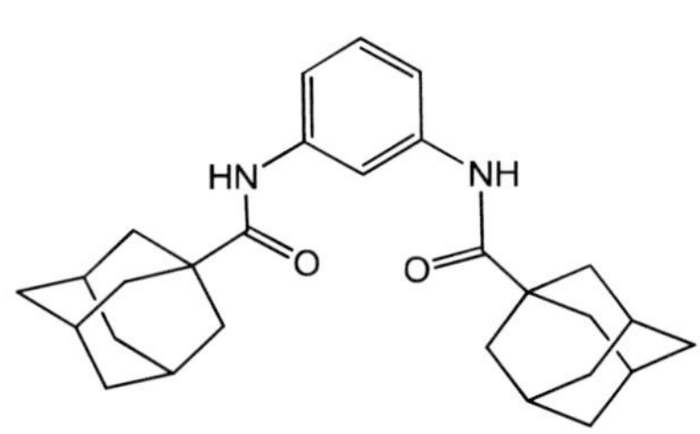

EDTA IN SUPRAMOLECULAR CHEMISTRY

Natural cyclodextrins are cyclic oligomers built up from 6, 7 or 8 glucopyranose units (named α, β **28**, and γCD, respectively) linked by α-(1-4)-glycosidic bonds. They have many practical applications in pharmacy and engineering [77-78] as they can form inclusion compounds by entrapping organic compounds (guests, G) in the hydrophobic cavity of the cyclodextrin (host, H). Their structures are very well known [79] .

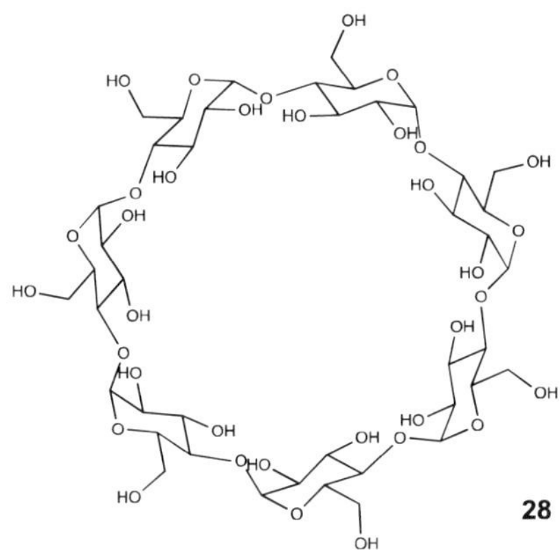

28

The complexation of adamantane derivatives with βCD has been the subject of numerous studies [80] (and references therein). The adamantyl group almost has a spherical shape which perfectly fits inside the cavity of βCD. Consequently the inclusion complex (stoichiometry H:G=1:1) is very stable, the equilibrium constant being among the strongest in the business. High equilibrium constant values is a prerequisite in order to obtain supramolecular polymers (see below) with a high degree of polymerization [81].

The following scheme illustrates the formation of a 1:1(H:G) complex:

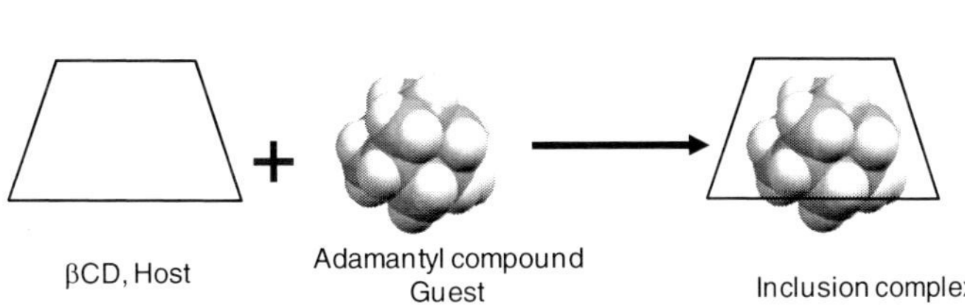

The synthesis of metal chelant cyclodextrin derivatives was described by Capretta et al. [82]. The chelant moieties were attached to C6 of a pyranose unit via a short alkyl spacer and by reacting EDTA monoanhydride **3** with amino derivatives of cyclodextrins yielding the monoamide **4.4** According to these authors ^{51}Cr can effectively label the chelate-cyclodextrin derivative, and allowing its progress through the intestine to be monitored, although no other data have been provide.

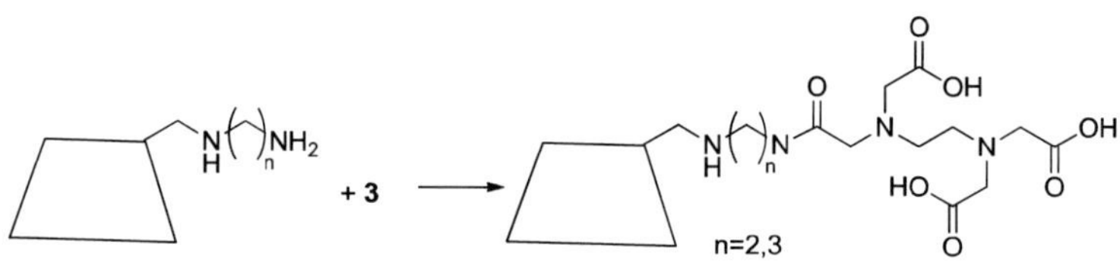

4.4

With a similar purpose these authors prepared **4.5** for chelating ^{99m}Tc.

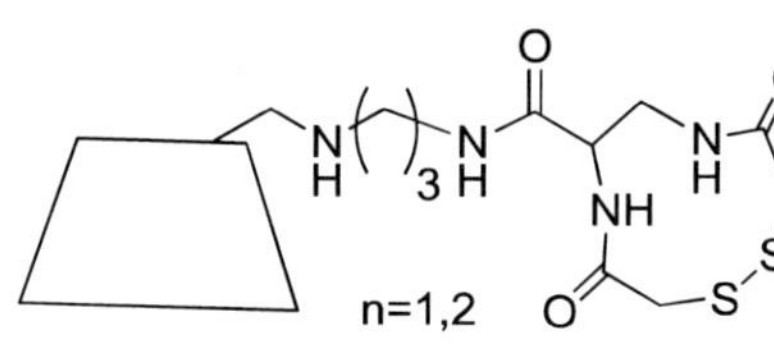

4.5

EDTA can be used to synthesize βCD dimers. For this purpose a monoamino derivative of βCD is previously prepared. The substitution can be performed by any of the two rims of the cyclodextrin. Yan et al. [83] have prepared βCD derivatives with an amino group at C6 (located at the primary rim), C3, or C2 (located at the secondary rim) carbon atoms of a glucose unit, and each of these derivatives was linked to EDTA, forming bis(amides). The EDTA spacer preserves its ability of forming chelates with metal ions as Ce(IV) and Zn(II) and the cyclodextrin residues preserve the property of forming inclusion complexes with organic molecules. With fluorescent guests as **29** and **30**, the equilibrium constants are high (10^4-10^6 $mol^{-1}dm^3$). More important is that this binding capacity is also common to the chelates. The Ce(IV) chelates were used as catalysts for the hydrolysis of the phosphodiester **31**. The kinetics follows a Michaelis-Menten mechanism evidencing a cooperative effect between the βCD cavities and the catalytic Ce(IV) chelate complex. **32** shows a schematic presentation for the cooperation of the hydrophobic binding with the catalytic ion group.

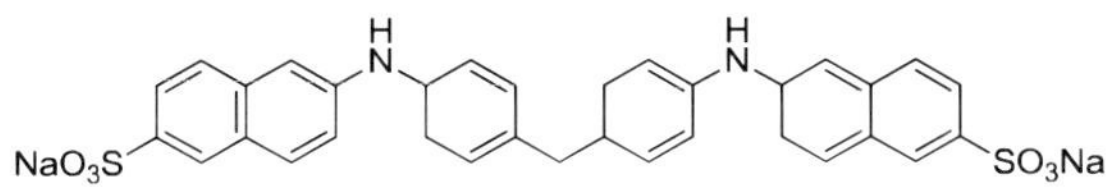

29

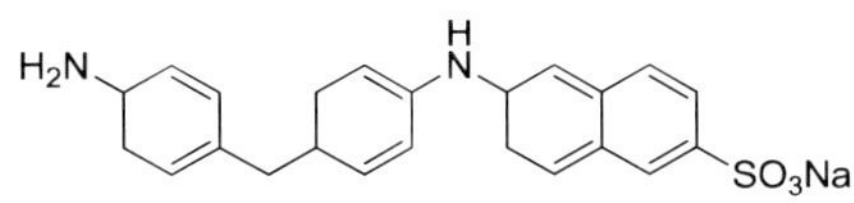

30

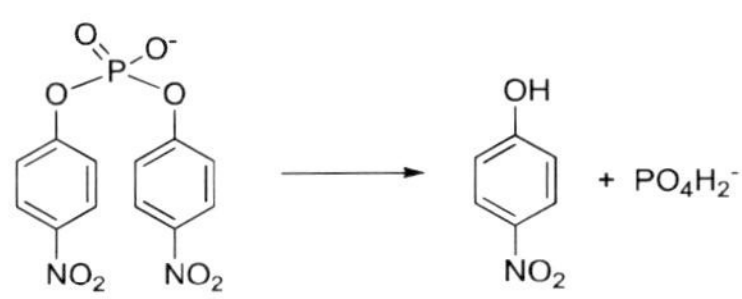

31

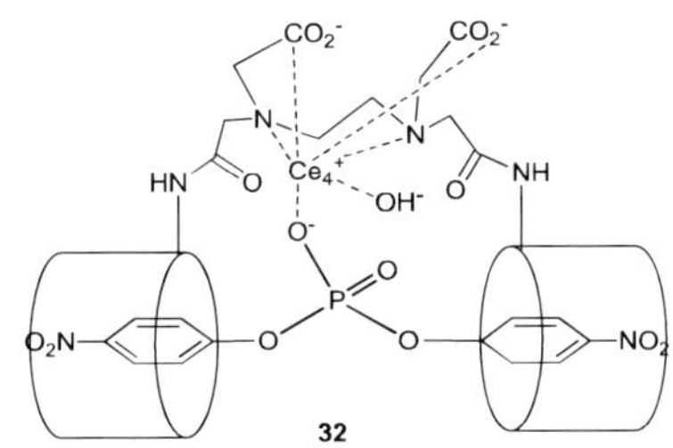

32

A EDTA-βCD dimer **5.5** was synthesized from EDTA dianhydride and a propylamino (appended on C2 of the glucose residue)-βCD derivative by Michels et al. [84]. *p-tert*-butylbenzoate, bound by the dimer cavities, forming a 1:2 (H:G) complex with an association constant of 10^4 $mol^{-1}dm^3$ in water, acts as a sensitizer for Eu(III) and Tb(III).

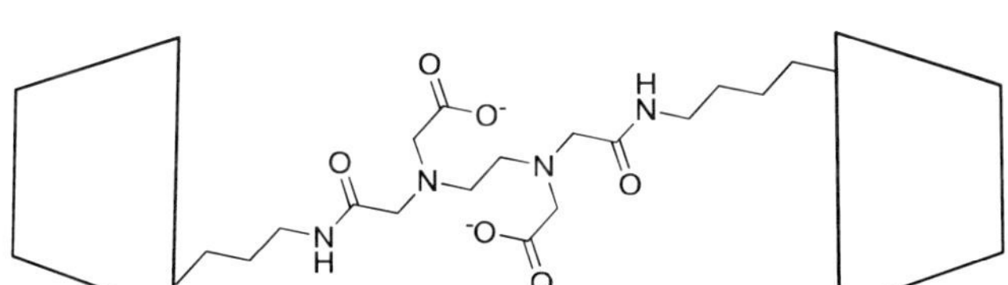

5.5

The Eu(III) complex of the cyclodextrin dimer forms strong 1:1 complexes (K ~ 10^7 $mol^{-1}dm^3$) with tailor made bis(propylamido adamantyl)-functionalized biphenyl sensitizers **33** and **34** in water, both cyclodextrin residues being involved in the binding of these guests. Amides **33** and **34** act as sensitizers for Eu(III) luminescence. The cyclodextrin residues increase the effective concentration of the sensitizer in the vicinity of the Ln(III) center. It is to say, for these biphenyl-based antennae both binding to βCD and coordination to the Ln(III) center are crucial for efficient sensitization.

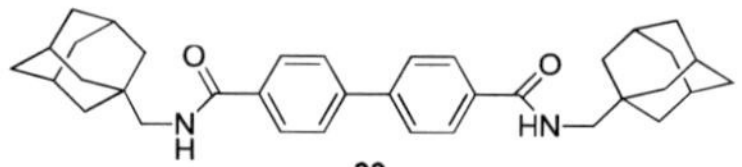

33

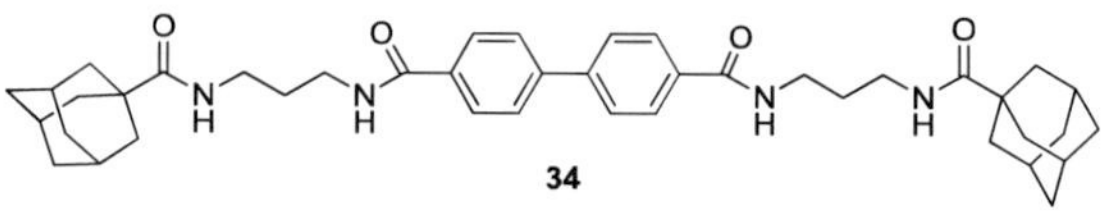

34

Analogously, the EDTA-tethered CD dimer **5.6** was obtained by Mulder et al. [85]. For the synthesis of **5.6** a similar procedure than for **5.5** was followed but sugar units of the CDs of **5.6** are altrosidic. Either (partial) protonation or metal chelation of the EDTA tether were used to obtain a series of structurally similar CD dimers with different tether charges and flexibilities. For instance, addition of $CaCl_2$ to aqueous solutions of **5.5** or **5.6** gave the neutral complexes, while the addition of $EuCl_3$ gives positively charged complexes. The organization around the metal ion of the EDTA bridge restricts the flexibility of the metal chelated cyclodextrin dimers in relation to the uncomplexing metal cyclodextrin dimers. Partial protonation modifies the charge of the tether but the flexibility essentially remains unmodified. Complex formation between two charged porphyrin guests by previous chelated and unchelated dimers was studied by isothermal titration calorimetry (ITC). The complexes have different stoichiometries. By combining chelated, unchelated dimers, partially protonated dimers, and the two guests, twelve inclusion complexes were studied. It has been shown that both the tether charge and the flexibility of the CD dimers strongly influenced the binding properties, giving rise to binding constants differing by factors up to 22.

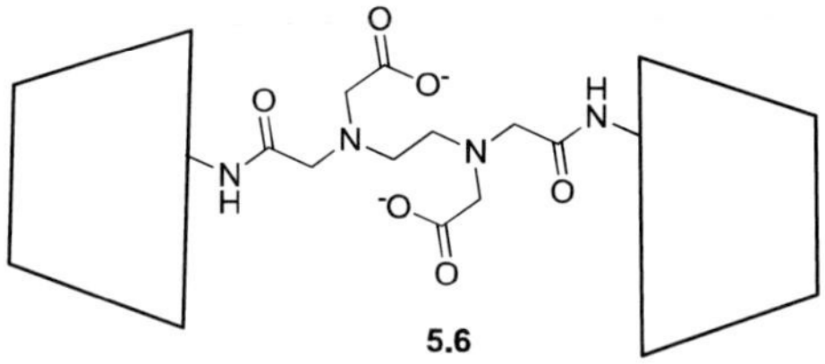

5.6

Although EDTA has been used to built supramolecular ionic networks [86], some of its derivatives have been more extensively used for. Thus when the adamantyl dimer **5.4** reacts with βCD the molecular stoichiometry of the complex should be 2:1 (H:G), the experimental value, determined from ITC experiments being 1.91(±0.01):1 [72].

The βCD dimer **5.7** (obtained by reacting a C6-amino βCD derivative with **2**) can include two adamantyl monomers to form an inclusion complex with a 1:2 (H:G) stochiometry [72]. The experimental average value for four adamantyl guests was 1:1.96(±0.04). Within experimental error, the two cyclodextrin residues in the dimer behave independently to each other when complexing the adamantyl derivatives [87].

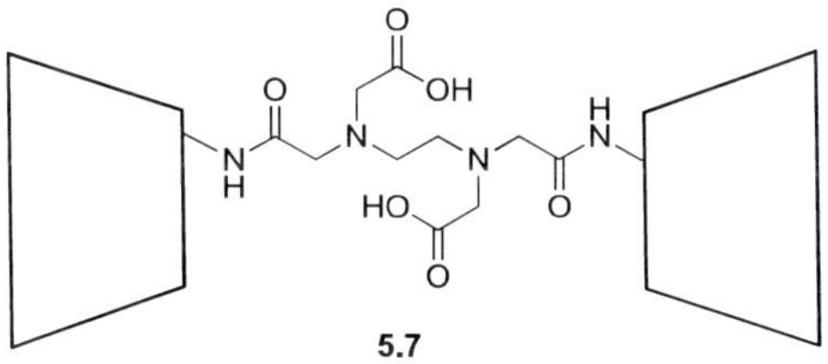

5.7

The most interesting case is when a βCD dimer and an adamantyl dimer react since a supramolecular linear polymer **35** is formed [72,88]. The complex formation of dimer **5.4** by some mono 6-amino-βCD dimers was also studied by Hocquelet et al. [89] by ITC and NMR techniques.

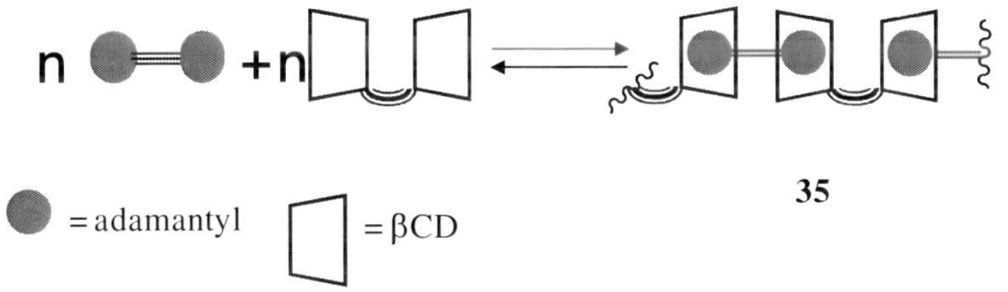

35

Dimer **5.5** was also used by Mulder et al. [90] to complex a bis(adamantyl)-functionalized calix[4]arene in solution. The binding constant (= 1.2×10^7 $mol^{-1}dm^3$) is orders of magnitude higher than the intrinsic binding constant for a single βCD-adamantane interaction. The strongly negative entropy of binding is attributed to restriction of mobility for both guest and host caused by the divalent interaction.

The formation of supramolecular polymers can be further pursued by increasing the number of binding sites of either the host or guest and in the case of a βCD trimer with **5.4** a Cayley tree structure **36** was observed [72]. A similar structure was obtained from the same βCD trimer and **20.2** (in sodium salt form) [91].

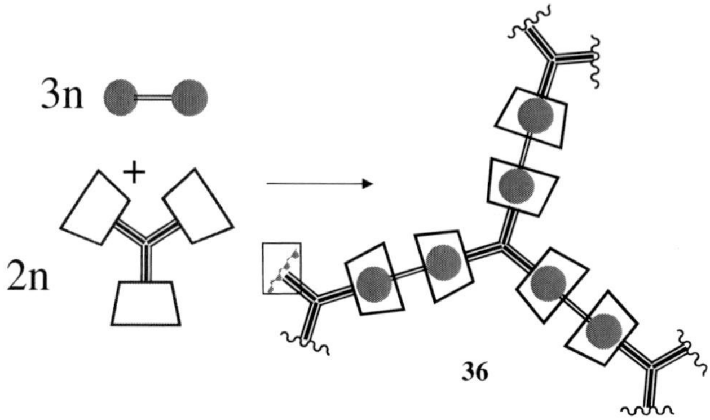

36

The hydrophobic guest and the cyclodextrin host can be residues of the same monomer. For instance, this is the case of the ditopic monomer **37** which forms the polymer **38** [92].

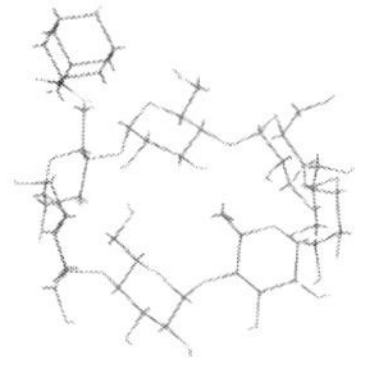

37

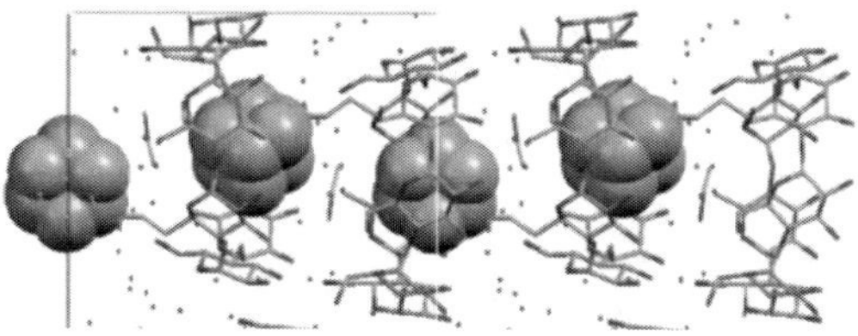

38

EDTA can be used as a bridge to form a self-aggregation unimer **7.1** carrying two host binding sites of βCD and two hydrophobic guest binding sites (two *p-tert*-butylphenyl moities) [93]. Static and dynamic light scattering measurements evidence the formation of supramolecular entities in aqueous solution with a low degree of polymerization (<10) and NMR experiments evidence that the *p-tert*-butylphenyl group is located at three different environments. Two of these sites correspond to the typical locations of a guest inside the host cavity and in the bulk solution, while the third one corresponds to a *p-tert*-butylphenyl group sandwiched between the outer surfaces of at least two cyclodextrin residues, forming an external complex. **7.1** was obtained by a two steps procedure. First, a diamide was obtained by reacting EDTA dianhydride and *p-tert*-butylphenylamine, followed by the formation of the diamide with 6-amino-βCD in the presence of DIC/HOBT.

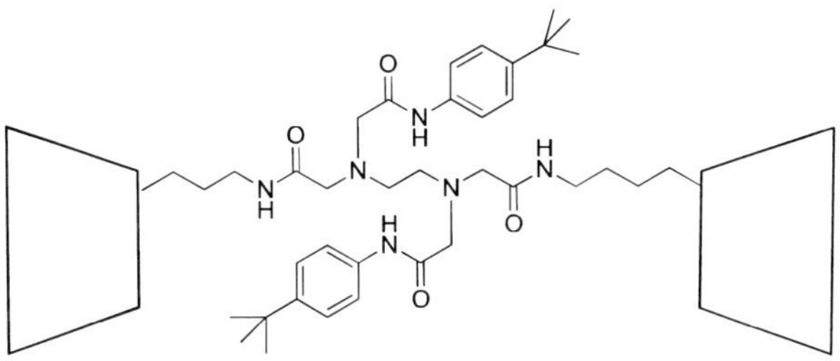

7.1

EDTA tetrasubstituted conjugates with two types of residues can be projected for other purposes. For instance Ottenbrite and Hu [94] have obtained **7.2** as possible oral drug delivery systems. As for **7.1** the synthesis is a two steps procedure: a first step in which EDTA dianhydride is open, and a second one implying the ester formation with poly(ethylene glycol), PEG, by activation of the carboxylic acid. Other polymer derivatives are commented on below.

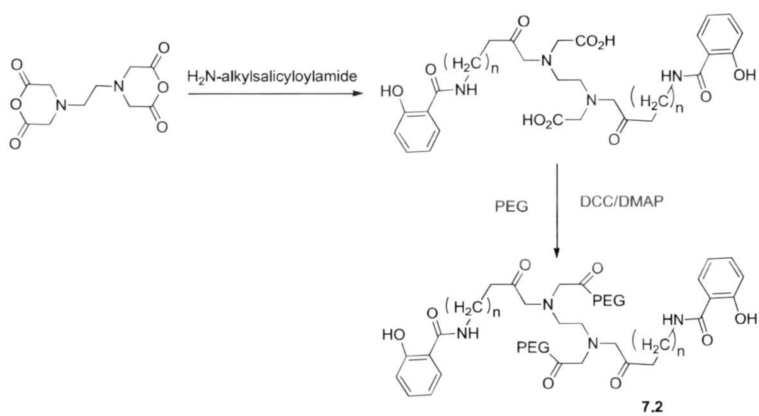

7.2

Cholesterol **39** is a lipid found in the cell membranes which plays a central role in many biochemical processes, such as regulating membrane fluidity. Cholesterol is highly hydrophobic and its solubility in water is very low.

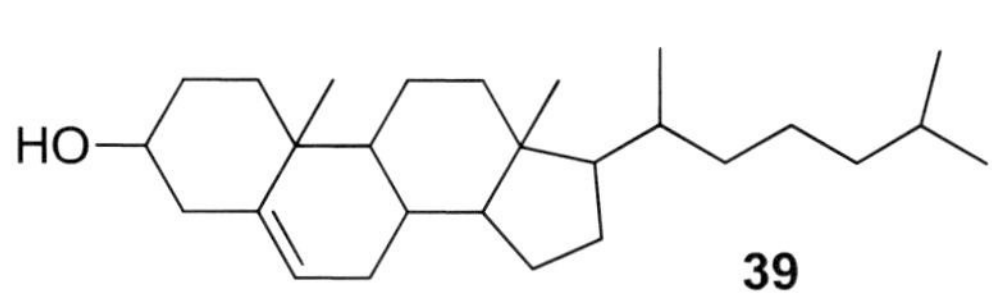

39

There are several reasons which justify the interest on cholesterol/ cyclodextrin interactions all over the years. They have been recently commented on by Alvarez et al. [95] and, among them, we can mention that cyclodextrins can be used to obtain low cholesterol foods even at industrial scale, βCD being the most frequently host used for that purpose. The subject has been partially reviewed by Jiang et al. [96] (see also [95] and [97] for references). However, the cholesterol-βCD inclusion complex has a limited aqueous solubility, but availability of solubilized cholesterol in aqueous solutions has obvious advantages. In this context, it is important to remark that the cholesterol length is almost twice the height of the cyclodextrin cone and **5.7** dimer, being highly soluble in water, is an obvious choice for enhancing the solubility of cholesterol. Alvarez et al. have shown that the phase-solubility diagram exhibits an A_L type profile, meaning that the solubilized cholesterol increases linearly with the concentration of the dimer, reaching concentrations up to 6×10^{-3} mol dm^{-3} in the presence of 0.025 mol dm^{-3} of **5.7**. The cholesterol-dimer complex has a 1:1 stoichiometry, the equilibrium constant being K=(8.8±0.2)$\times10^{4}$ mol^{-1} dm^{3}. This nice application of EDTA-cyclodextrin derivative can be easily extended to other low soluble drugs and hydrophobic organic molecules.

Other derivatives in which half of the EDTA structure is bonded to C6 in cyclodextrins have also been obtained [98]. Complexes with Gd(III), which could be useful as contrast agents in MRI were studied.

BIS(IMIDE) DERIVATIVES (6)

Compounds of the **6** family were patented by Dazzi [99] as intermediates for pharmaceutical, textile aids, and hardening agents for epoxy resins. Other derivatives have been synthesized by Takemitsu et al. [42], Herbst and Eckert [100] and Woo [101].

The compound **6.1** (R= C_8H_{17} in **6**), obtained by Takemitsu et al. [42], has good dispersing power for TiO_2 pigments in organic solvents. However no chelating properties have been reported, probably, because the lack of mobility of the carboxylic residues prevents the formation of stable chelates. In fact, Takemitsu et al. have hydrolyzed the bis(imides) to prepare bis(amides) (starting amines: hexyl, octyl, nonyl, decyl, undecyl, and dodecyl) which were dissolved in EtOH, kept at pH 6-7, and mixed with aqueous solutions of metal nitrates to prepare the chelates. The chelates (yields in the range 54-92%) were soluble in some organic solvents, but did not dissolve in water.

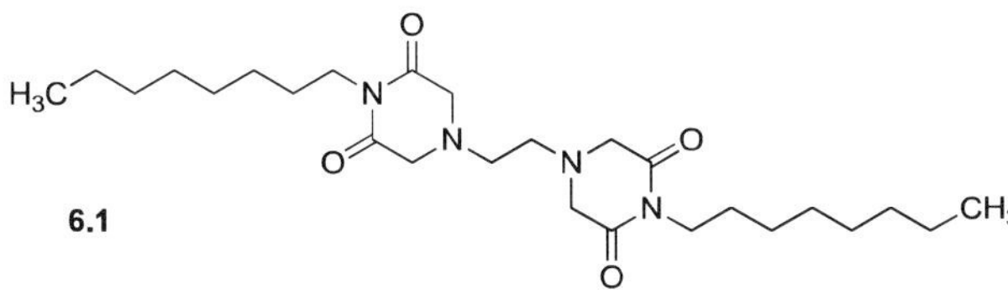

6.1

NN-Disubstituted derivatives of the bis(imide) of *NN*-ethylenebis (iminodiacetic acid) were prepared by alkylation of the disodium derivative of the bis(imide) [102]. E.g., treatment of **6.2** with NaOMe in refluxing MeOH followed by $BrCH_2CO_2Me$ in DMF gave 41% of **6.3**. The hydrolysis and methanolysis of the bis(imide) and some of its *N,N'*-disubstituted derivatives were promoted by Cu(II) and afforded copper complexes from which the expected bis(half amides) of *N,N'*-ethylenebis(iminodiacetic acid) were obtained by the action of hydrogen sulphide. E.g., reaction of **6.2** with $Cu(OAc)_2$ in refluxing H_2O for 10 min gave 80% of the Cu complex **6.4**, which after treatment with H_2S in H_2O gave 57% of the bis(amide) **6.5**.

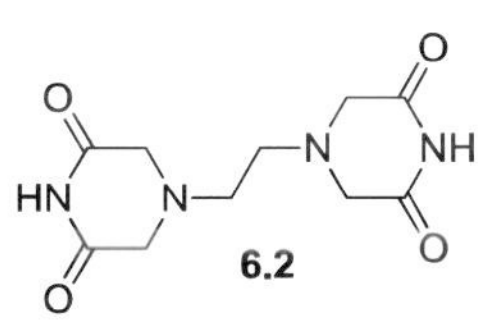

6.2

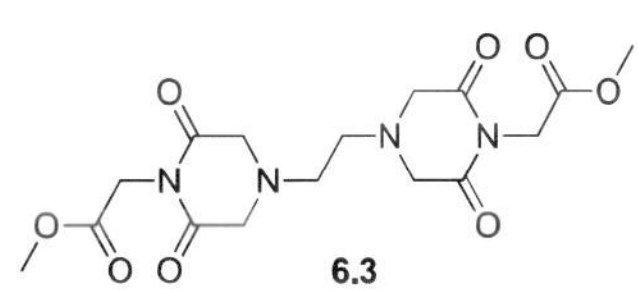

6.3

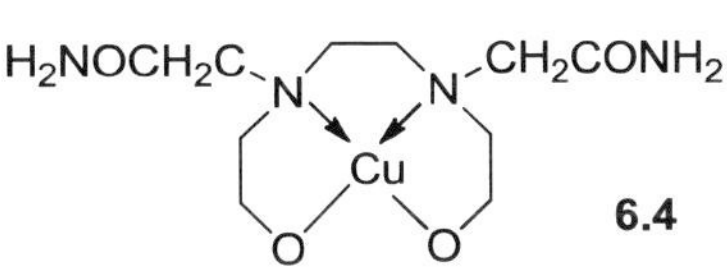

6.4

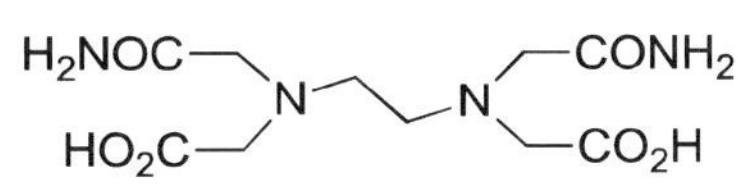

6.5

Previous compounds can be observed as piperazine derivatives, and other compounds have been published [103]. **6.6** was obtained from the reaction of dodecyl amine and EDTA [100] which was treated with Me_2SO_4 in butanol to give 1,2-bis(N-dodecyl-2,6-dioxo-4-piperazinyl)ethane dimethyl sulfate monoquaternary salt. The products were proposed for the formulation of fabric softeners for softening washed laundry.

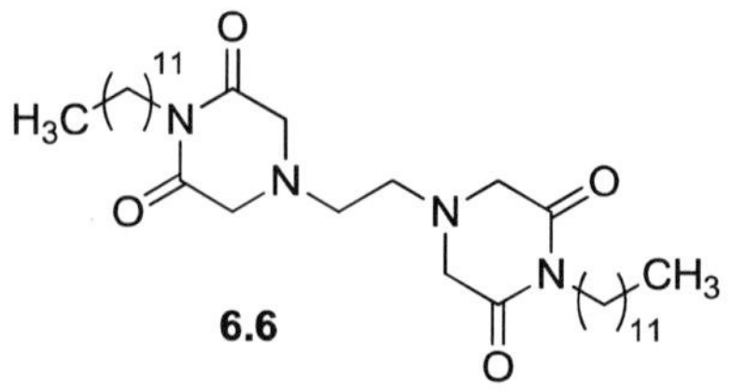

6.6

Some dicarboxylic acid derivatives were obtained by refluxing EDTA dianhydride with alkyl amino acids. For instance, the 6-aminohexanoic acid gives the **6.7** derivative. These diacid derivatives reacted with 1,6-hexanediamine to give a nylon salt and heating this salt with nylon 66, in the presence of $Mn(H_2PO_2)_2.H_2O$, gave the copolyamide [101] .

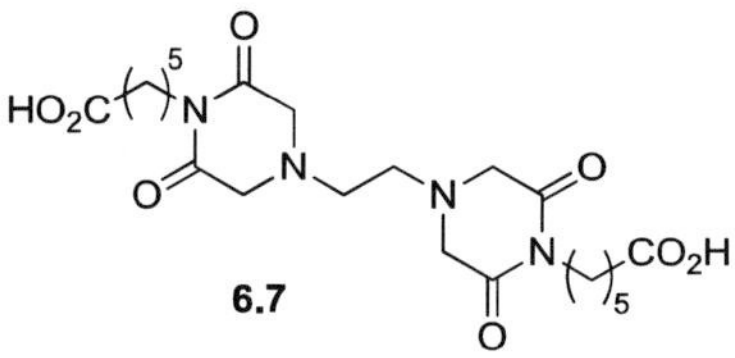

6.7

PROTEINS AND POLYMERS

The R group in **8** can be a polymer [104]. EDTA was introduced to glycidyl groups attached to 1:1 and 2:1 ethyl acrylate-glycidyl methacrylate copolymer, 1:1 butyl acrylate-glycidyl methacrylate copolymer, and poly(glycidyl methacrylate) to form polymers **8.2** containing chelating ligands as pendant groups in good yields (37-79%). Chelate polymers were obtained by the reaction of the polymers with some transition metal ions (Ni(II), Cu(II), Zn(II)). Some of the chelate polymers have been tested as surfactants and others had excellent emulsifying capacities for kerosene-water mixtures.

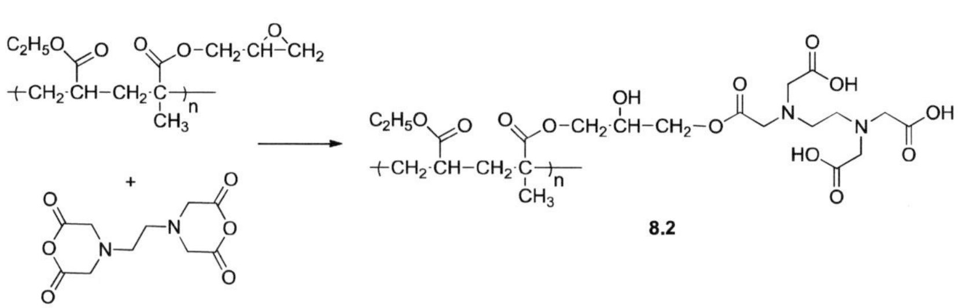

8.2

The R group carrying the amino group in the family of compounds **5**, can be a protein. Hwang et al. [105] modified the lysyl residues of soy protein isolate (SPI) with EDTA dianhydride, which was later cross-linked with glutaraldehyde. As in other analyzed R residues, either a monoamide or bis(amide) may be formed. In the later case high molecular weight polymers are formed while in the former case three carboxyl moieties per lysyl residue are incorporated into the protein. The authors confirm that the reaction essentially leads to the formation of bis(amides). However, when SPI is first partially denatured (at high pH and high temperature), the main reaction product is the monoamide. The native SPI is predominantly a β-type protein (it contains about 60% β-sheet). The reaction with EDTA dianhydride decreases the β-sheet content and increases both β-turn and random-coil contents, showing that although the structure of SPI is altered, a significant amount of β-sheet structure is retained. The cross-linked polymeric gel, prepared by the reaction of modified SPI with aqueous glutaraldehyde, results in the formation of a gel able to take up about 105 g of water/g of gel.

EDTA-based metal chelates that can be attached to the cysteine residue of a protein via a disulfide bond have been described [41]. The complexes were used after coordination of lanthanides to align trigger factor and apo-calmodulin in solution.

Single-phase $Y_1Ba_2Cu_3O_{7-x}$ was synthesized by Maeda et al. [106] by a polymer-metal chelate precursor technique. For this purpose EDTA-ethylenediamine copolymer polyamide **5.8** was used as chelating polymer for Y, Ba, and Cu.

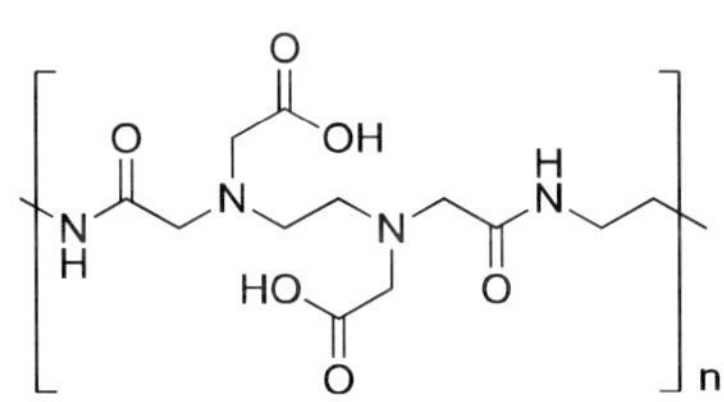

5.8

The ability of the polymer to form chelates was confirmed by titration curves of the polymer in the absence and presence of each metal ion. The polymer (no metal added) behaves as a dibasic compound since two equivalent points (at pH 5.5 and 9.0), corresponding to the neutralization of the free carboxylic and amino groups, were observed.

The polymer was dissolved in water and the nitrates of Y, Ba, and Cu (1:2:3 in mole ratio) were stoichiometrically added forming a homogenous solution. Removal of the solvent leaves the polymer with metal ions without phase separation or crystallization of any of the metal compounds. This precursor was dried, calcined, and sintered to develop the superconducting (1,2,3) phase which, according to X-ray pattern, was pure and had good superconducting properties, the onset being observed at around 90.0 K and the resistance reaching substantially zero at 88.6 K. When comparing the imide and bis(amides), it is obvious that the polymer only preserves the chelating ability of EDTA when at least two carboxylic groups are not involved in the formation of covalent amide bonds.

Closely related with previous work is the paper published by Naka et al. [107] in which the ethylenediamine in EDTA is substituted by 1,2-diaminocyclohexane. The corresponding dianhydride was open by several diamines to prepare the polymer. The polymer obtained with 1,2-diaminocyclohexane was poured into tetrahydrofurane with 1/2 equivalent molar amount of metal nitrates of Y, Ba, and Cu (1:2:3 in mole ratio) to precipitate a polymer–metal chelate. This precursor was used to obtain $YBa_2Cu_3O_{7-x}$ superconducting films.

Imai [108] have use the dianhydride **2** to react with a diamine to obtain poly-2,6-piperazidendione polymers **6.8**. A typical synthetic route is illustrated below with the diamine bis(4-aminophenyl) ether.

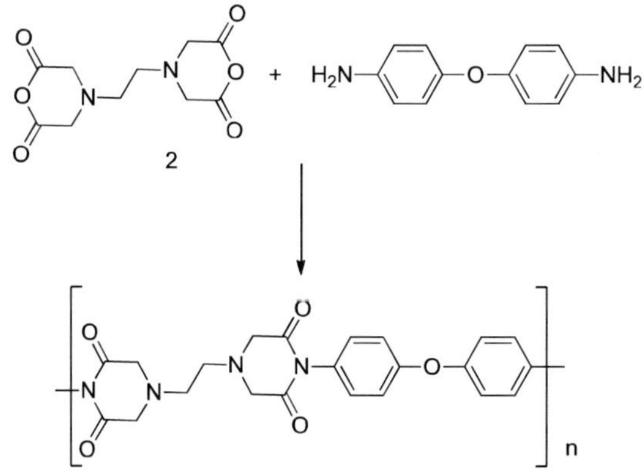

6.8

The polymer is formed in two steps. The ring-opening polyaddition reaction yields a high molecular weight polyamide-acid in the first step and subsequent dehydration of this precursor gives the final poly-2,6-

piperazinedione. In the first step of polymerization, solvents as dimethylformamide, dimethylacetamide, and N-methyl-2-pyrrolidone were used. In the second step, cyclodehydration to poly-2.6-piperazinedione was carried out by refluxing polyamide acid film in an excess of acetic anhydride and drying. The poly-2,6-piperazinedione thus obtained did not melt at a temperature below 360 °C, and was insoluble in all organic solvents tried.

Prior to polymer synthesis, Imai prepared, as model compounds, the bis(amide) **5.9** and bis(imide) **6.9** derivatives of aniline.

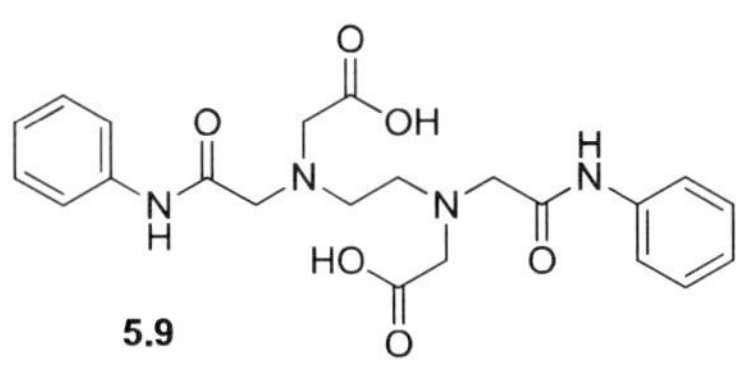

5.9

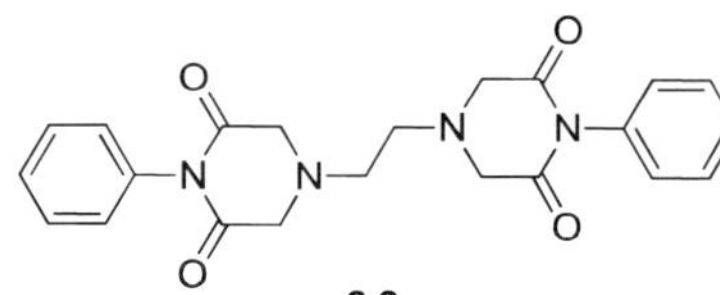

6.9

Polyesters with pendant carboxylic groups were synthesized by polycondensation of dianhydride **2** and diethylenetriaminepentaacetic acid dianhydride with PEGs of different chain-lengths. The polymeric products were soluble in water, acetone, chloroform and toluene. The linear polyesters are able to bind metal ions in aqueous solutions, and this property was found to be strongly pH dependent. The addition of Cu(II) to **9.3** solutions led to a color change from blue to green; however, no precipitation was observed.

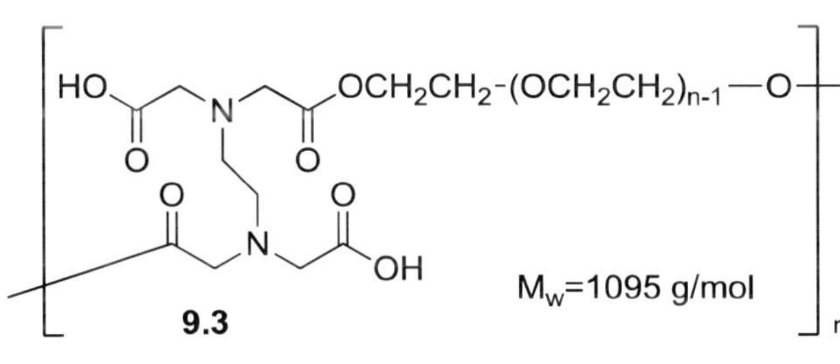

9.3

In a series of papers Di Noto et al. [109-111] have synthesized several polymer electrolytes **9.4** and **9.5** from PEG and EDTA.

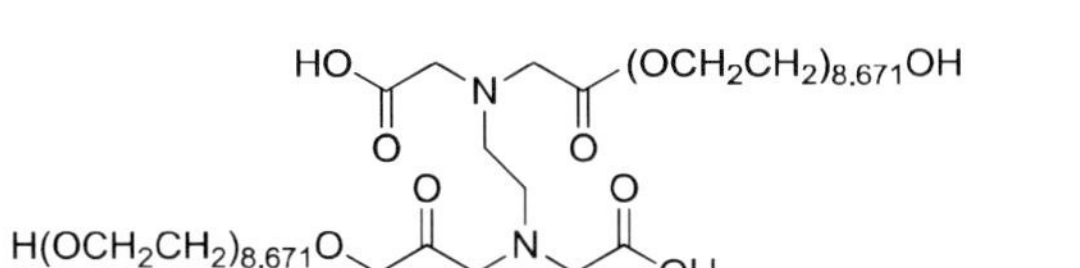

9.4

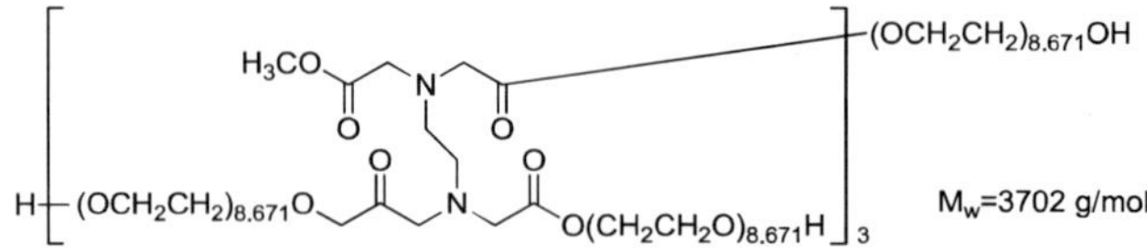

9.5

From these polymers, ionic complexes with LiCl and $MgCl_2$ were formed: **9.5**/$(LiCl)_{6.39}$, **9.5**/$(MgCl_2)_{8.23}$, **9.5**/$(LiCl)_{6.39}$/$(MgCl_2)_{6.16}$, **9.4**/$(LiCl)_{2.26}$, **9.4**/$(MgCl_2)_{1.98}$, **9.4**/$(LiCl)_{2.26}$/($MgCl_2)_{1.98}$. The compounds were studied by impedance spectroscopy at 20 Hz to 1 MHz and different temperatures. The studies demonstrated that the EDTA coordination site is selective toward magnesium ion species. This strong coordination inhibits the conductivity of these by reducing the concentration of magnesium cation species able to migrate and by decreasing the segmental mobility of host materials owing to the formation of magnesium-ion coordination cross-links between different host polymer chains. Lithium ions are the most mobile cations in these systems, the ion preferentially interacting with both of the oxygen atoms of the EDTA sites and polyether moieties.

Kahana et al. [112] proposed a conceptual approach for the synthesis of 6-carboxy and 6-amino polyamide-linked EDTA, AE_n. **40** is an example of AE_2.

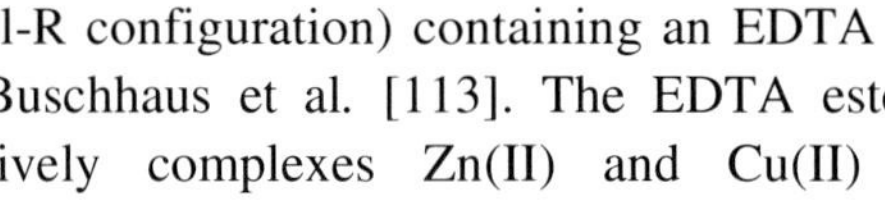

40

The synthesis and metal complexation of chiral depsipeptide dendrimers **41** and **42** (with all-R configuration) containing an EDTA ester derived core was reported by Buschhaus et al. [113]. The EDTA ester cavity of these dendrimers selectively complexes Zn(II) and Cu(II) ions leading to diastereoselective folding. As a model case, the coordination behavior of the tetramethyl ester of EDTA **10.1** of those ions was also studied. Stable metal complexes were formed upon stirring of dendrimers **41** and **42** in $CHCl_3$ at room temperature in the presence of a large excess of Zn(II) and Cu(II) salts. The dendrimers **41**-$ZnCl_2$ and **42**-$ZnCl_2$ were isolated as colorless solids, whereas the copper dendrimers **41**-$CuCl_2$ and **42**-$CuCl_2$ were obtained as light green solids. All these metallodendrimers were found to be very soluble in apolar and polar aprotic solvents such as CH_2Cl_2, $CHCl_3$, and CH_3CN, and soluble to some extent in polar protic solvents such as CH_3OH and CH_3CH_2OH. **43** and **44** are the chelating motifs of both dendrimers.

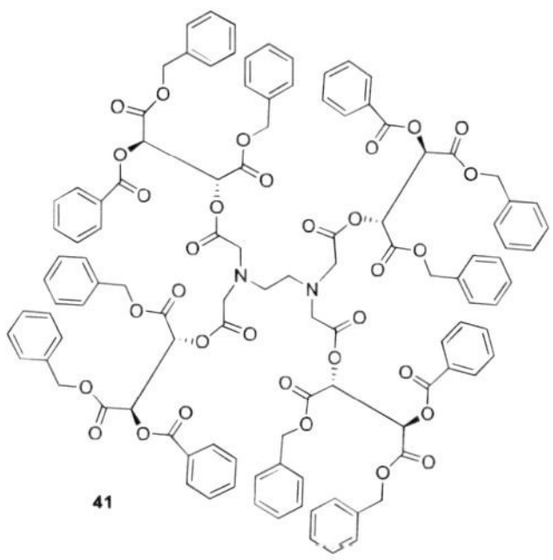

41

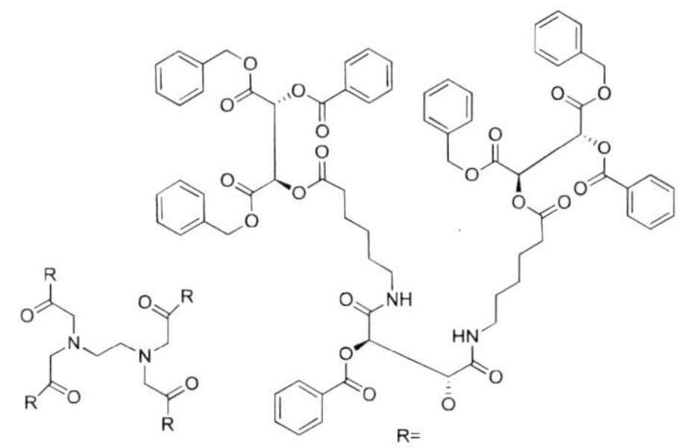

42

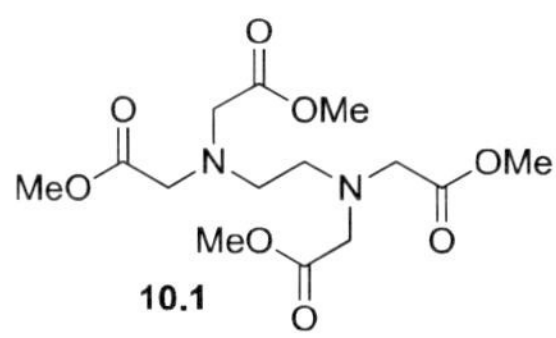

10.1

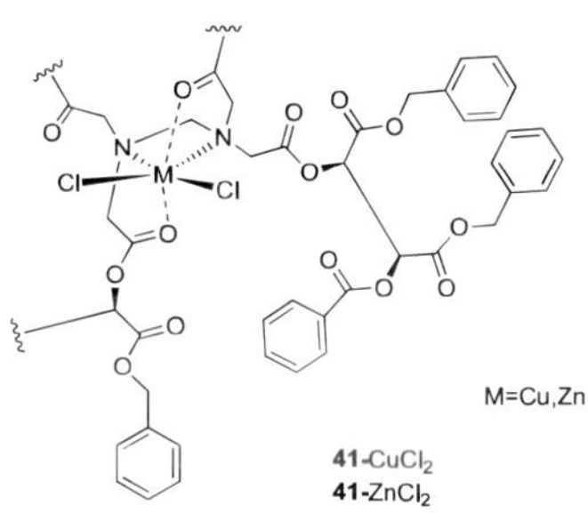

43

41-$CuCl_2$
41-$ZnCl_2$

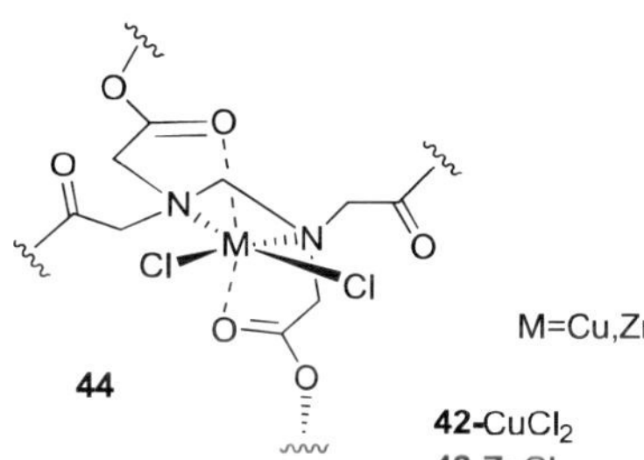

44

42-$CuCl_2$
42-$ZnCl_2$

The crystal structures of both model cores **10.1**-$ZnCl_2$ and **10.1**-$CuCl_2$ show that they exclusively adopt the cis-α configuration. On the basis of calculation of the CD spectrum of Δ*cis*-α(*S*,*S*)-**10.1**-$CuCl_2$, the absolute configurations of the preferentially formed diastereoisomers of the related chiral dendritic complexes could be assigned as Λ*cis*-α(R,R)(all-R_{dend})-**41**-$CuCl_2$ and Δ*cis*-α(*S*,*S*)(all-R_{dend})-**42**-$CuCl_2$.

Final Remarks and Perspectives

The examples reviewed above clearly show symbiotic effects between the characteristics of EDTA and those of the bonded moieties, as the new derivatives exhibit properties that both residues have independently. In some cases, the simultaneity of the two moieties in the same compound gives rise to new properties which are nonoccurrence in the compounds independently considered or in their mixtures in solution. Occasionally synergic effects are also evident. Chelating surfactants and polymers, polyelectrolytes, enhancement of the solubility in water of hydrophobic compounds, mimicry of enzymes, scissor molecules, etc. are illustrative examples of these phenomena.

Some published compounds are far from being fully characterized. For example, size, aggregation number, or fraction of counterions of the new chelating surfactants have not been determined. Biological properties of EDTA/natural product (bile acids are examples) derivatives are almost unexplored, and many other compounds, with tailored properties, have not been synthesized. This is particularly right for asymmetric derivatives in which symbiosis and synergic effects between at least three units can be designed.

ACKNOWLEDGMENTS

The authors thank the Ministerio de Ciencia y Tecnología, Spain, Project MAT2010-61721 for financial support.

REFERENCES

[1] Flaschka, H. A. (1964) *EDTA Titrations, (Pergamon Press, Oxford*

[2] Lippard, S. J. Metals in Medicine, *Bioinorganic Chemistry chapter 9.*

[3] Choppin, G. R. & Schaab, K. M. (1996). Lanthanide(III) complexation with ligands as possible contrast enhancing agents for MRI, *Inorg. Chim. Acta*, *252*, 299-310.

[4] Griko, Y. V. (1999). Energetics of Ca^{2+}- EDTA interactions: calorimetric study, *Biophys. Chem.*, *79*, 117-27.

[5] Claudio, E. S., Ter Horst, M. A., Forde, C. E., Stern, C. L., Zart, M. K. & Godwin, H. A. (2000). 207Pb-1H Two-Dimensional NMR Spectroscopy: A Useful New Tool for Probing Lead(II) Coordination Chemistry, *Inorg. Chem.*, *39*, 1391-7.

[6] Fisher, A. E. O. & Naughton, D. P. (2004). Synthesis, characterisation and ferric ion complexation studies of chelating probes for oxidative stress, *Transition Metal Chem.*, *29*, 315-9.

[7] Platas-Iglesias, C., Corsi, D. M., Vander Elst, L., Muller, R. N., Imbert, D., Buenzli, J.-C. G., Toth, E., Maschmeyer, T. & Peters, J. A. (2003). Stability, structure and dynamics of cationic lanthanide(III) complexes of N,N'-bis(propylamide)ethylenediamine-N,N'-diacetic acid, *Dalton Trans.*, 727-37.

[8] Wang, Y.-M., Wang, Y.-J. & Wu, Y.-L. (1999). Synthesis, stability and nuclear magnetic resonance studies for the complexation of trivalent lanthanides, Ca^{2+}, Cu^{2+} and Zn^{2+} by N,N'-bis(amide) derivatives of ethylenedinitrilotetraacetic acid, *Polyhedron*, *18*, 109-17.

[9] Esterhuysen, C., Bredenkamp, M. W., van Kralingen, L., Adendorff, H. J., Jacobs, E. P. & Swart, P. (2006). Aqua[3,6-bis (methoxycarbonylmethyl)-3,6-diazaoctanedioato]copper(II) dihydrate, *Acta Crystal. E62*, m355-7.

[10] Leonov, A., Voigt, B., Rodriguez-Castaneda, F., Sakhaii, P. & Griesinger, C. (2005). Convenient synthesis of multifunctional EDTA-based chiral metal chelates substituted with an S-mesylcysteine, *Chem. Eur. J.*, *11*, 3342-8.

[11] Zeman, I. (1980). Analytical investigation of the chelating surfactant, *Tensioactivos: Biodegradabilidad, Fis.-Quim. Apl., Jorn. Com. Esp. Deterg.*, *11th*, 553-9.

[12] Micklitsch, C. M., Yu, Q. & Schneider, J. P. (2006). Unnatural multidentate metal ligating α-amino acids, *Tetrahedron Lett.*, *47*, 6277-80.

[13] Buica, G. O., Bucher, C., Moutet, J.-C., Royal, G., Saint-Aman, E. & Ungureanu, E. M. (2009). Voltammetric sensing of mercury and copper cations at poly(EDTA-like) film modified electrode, *Electroanalysis*, *21*, 77-86.

[14] Kaluderovic, G. N., Schmidt, H., Wagner, C., Merzweiler, K. & Steinborn, D. (2007). Synthesis, characterization and crystal structure of palladium(II) complexes containing EDTA tetraalkyl ester ligands, *Collect. Czech. Chem. Commun.*, *72*, 560-58.

[15] Furuta, T., Sakai, M., Hayashi, H., Asakawa, T., Kataoka, F., Fujii, S., Suzuki, T., Suzuki, Y., Tanaka, K., Fishkin, N. & al, e. (2005). Design and synthesis of artificial phospholipid for selective cleavage of integral membrane protein, *Chem. Commun.*, 4575-7.

[16] Takeshita, T., Shimohara, T.-A. & Maeda, S. (1982). Synthesis of EDTA-monoalkylamide chelates and evaluation of the surface-active properties, *J. Am. Oil Chem. Soc.*, *59*, 104-7.

[17] Jaeger, D. A., Peacock, M. F. & Bohle, D. S. (2003). A Surfactant Transition Metal Chelate, *Langmuir*, *19*, 4859-62.

[18] Zhang, T., Wu, Y.-X., Huang, X.-F., Liu, J.-M., Xia, B., Zhang, W.-H. & Qiu, R.-L. (2012). Simultaneous extraction of Cr(VI) and Cu(II) from humic acid with new synthesized EDTA derivatives, *Chemosphere*, *88*, 730-5.

[19] Jaeger, D. A., Jose, R., Mendoza, A. & Apkarian, R. P. (2007). Surfactant transition metal chelates, *Colloids Surfaces A*, *302*, 186-96.

[20] Takeshita, T., Wakebe, I. & Maeda, S. (1980). Synthesis of EDTA-monoalkyl ester chelates and evaluation of the surface active properties, *J. Am. Oil Chem. Soc.*, *57*, 430-4.

[21] Wasow, G., Novak, J., Prochazka, K., Zastrow, L., Haensel, E., Hesselbarth, M., Wolf, C. & Kaufmann, D. (1990). Surfactant combination, especially for laundering at low or moderate temperatures, *Patent Ger. (East)*, DD 280114 A1 19900627.

[22] Novak, J., Sochorova, K., Kralovicova, S., Slais, Z., Tolman, J. & Prochazka, K. (1986). Universal foaming cleanser, *Patent Czech.*, CS 228477 B1 19840514.

[23] Csuk, R., Barthel, A., Brezesinski, T. & Raschke, C. (2004). Synthesis of pathogen inactivating nucleic acid intercalators, *Eur. J. Med. Chem.*, *39*, 975-88.

[24] Csuk, R., Raschke, C., Goethe, G. & Reissmann, S. (2005). Synthesis of monomeric acridine derived nucleic acid intercalators, *Z. Naturforsch. B 60*, 83-8.

[25] He, M., Fair, M. J. and Massaro, M. (1998). Liquid compositions comprising EDTA-derived chelating surfactants, *Patent*, EP 834307 A2 19980408.

[26] Prochazka, K., Novak, J. & Kaufman, K. (1992). Storage-stable liquid laundry detergent containing enzymes, *Patent*, CS 275821 B6 19920318.

[27] Novak, J., Prochazka, K., Kaufman, K. & Mikulcova, D. (1992). Stabilization of enzymes in liquid and paste laundry detergents, *Patent*, CS 273853 B1 19910411.

[28] Bumbalova, A., Havranek, E., Harangozo, M., Krenek, P. & Lubkeova, S. (1991). Preparation and biodistribution of metastable technetium-99-HT18, *J. Radioanal. Nuclear Chem.*, *154*, 53-9.

[29] Novak, J., Lopata, V., Prochazka, K. & Majer, J. (1980). Surfactants with chelating effect, *Tensioactivos: Biodegradabilidad, Fis.-Quim. Apl., Jorn. Com. Esp. Deterg., 11th* 317-33.

[30] Danil de Namor, A. F. Z.-F., Katherine; Pacheco Tanaka & David Alfredo. (1998). Thermodynamics of protonation and complexation of EDTA derivatives and metal cations in water, *J. Chem. Soc. Faraday Trans.*, *94*, 3105-10.

[31] Prochazka, K., Novak, J., Simunek, J., Peterka, V. & Krob, V. (1974). Detergent containing a surfactant with a capability for activation, *Patent*, (1974), FR 2202935 A1 19740510.

[32] Prochazka, K., Novak, J., Simunek, J., Peterka, V. & Krob, V. (1976). Detergent containing surfactant with activating power, *Patent*, US 3989634 A 19761102.

[33] Havranek, E., Bumbalova, A., Krenek, P., Komova, M. & Belakova, M. (1994). Development of new radiopharmaceuticals-technetium aminopolycarboxylate complexes, *Pharmazie*, *49*, 369-370.

[34] Eckelman, W. C., Karesh, S. M. & Reba, R. C. (1975). New compounds. Fatty acid and long chain hydrocarbon derivatives containing a strong chelating agent, *J. Pharm. Sci.*, *64*, 704-6.

[35] Agency, I. A. E. (2009). Technetium-99m Radiopharmaceuticals: Status and Trends.

[36] Danil de Namor, A. F., Cardenas Garcia, J. D. & Bullock, J. I. (1995). Synthesis and thermodynamic characterization of cobalt(II), nickel(II) and copper(II) complexes of ethylenediamine-N,N,N',N'-tetraacetanilide, *Pure Appl. Chem.*, *67*, 1053-8.

[37] Santos, M. A., Gama, S., Gano, L. & Farkas, E. (2005). Bis(3-hydroxy-4-pyridinone)-EDTA derivative as a potential therapeutic Al-chelating agent. Synthesis, solution studies and biological assays, *J. Inorg. Biochem.*, *99*, 1845-52.

[38] Urizzi, P., Souchard, J.-P. & Nepveu, F. (1996). EDTA and DTPA Analogues of Dipalmitoylphosphatidylethanolamine as Lipophilic Chelating Agents for Metal Labeling of LDL, *Tetrahedron Lett.*, *37*, 4685-8.

[39] Takeshita, T., Maeda, S., Imayoshi, M., Mitsuo, Y. & Tokumura, H. (1982). Studies on coal-oil mixture (COM). Part 2. Effect of additives on sedimentation stability of COM. I *Sekiyu Gakkaishi*, *25*, 389-97.

[40] Takeshita, T., Maeda, S., Imayoshi, M., Mitsuo, Y., Katayama, Y. & Kamikawa, T. (1983). Studies on COM. Part 3. Effects of additives on sedimentation stability of COM, *Sekiyu Gakkaishi*, *26*, 384-9.

[41] Haberz, P., Rodriguez-Castaneda, F., Junker, J., Becker, S., Leonov, A. & Griesinger, C. (2006). Two New Chiral EDTA-Based Metal Chelates for Weak Alignment of Proteins in Solution, *Org. Lett.*, *8*, 1275-8.

[42] Takemitsu, K., Miyauchi, N., Maeda, S. & Takeshita, T. (1977). Studies on the utilization of metal chelates. X. Studies on the EDTA-bis(N-alkylamide) chelate surfactants, *Yukagaku*, *26*, 362-6.

[43] Takeshita, T. & Maeda, S. (1970) Utilization of metal chelates. IV. C-alkyl EDTA chelates: preparation and evaluation of surface and interfacial tensions and dispersant properties, *Yukagaku*, *19*, 984-93.

[44] Yamashita, K., Miyauchi, N., Maeda, S., & Takeshita, T. (1979). Studies on the utilization of metal chelates. XI. Preparation of EDTA-bis (N,N-dialkylamide) chelate and estimation of the surface active properties, *Yukagaku*, *28*, 552-6.

[45] Maeda, S. & Takeshita, T. (1990). Metal chelate surfactants, *Yukagaku*, *39*, 723-8.

[46] Goto, M., Miyata, T., Kubota, F. & Nakashio, F. (1992). Selective separation of rare earth metals with new oil-soluble complexing agents, *J. Chem. Eng. Jpn*, *25*, 349-51.

[47] Kubota, F., Miyake, Y., Goto, M., Furusaki, S., Hano, T. & Nakashio, F. (2000). A novel extraction and separation method of rare earth metals utilizing an oil-soluble complexing agent, *Solvent Extraction Res. Develop. Jpn*, *7*, 139-51.

[48] Babu, P., Sangeetha, N. M. & Maitra, U. (2006). Supramolecular chemistry of bile acid derivatives: formation of gels, *Macromol. Symp.*, *241*, 60-7.

[49] Davis, A. P. & Joos, J.-B. (2003). Steroids as organising elements in anion receptors, *Coord. Chem. Rev.*, *240*, 143-56.

[50] Davis, A. P. (2007). Bile acid scaffolds in supramolecular chemistry: the interplay of design and synthesis, *Molecules*, *12*, 2106-22.

[51] Nonappa & Maitra, U. (2008). Unlocking the potential of bile acids in synthesis, supramolecular/materials chemistry and nanoscience, *Org. Biomol. Chem.*, *6*, 657-69.

[52] Tamminen, J. & Kolehmainen, E. (2001). Bile Acids as Building Blocks of Supramolecular Hosts, *Molecules*, *6*, 21-46.

[53] Davis, A. P. (2006). Anion binding and transport by steroid-based receptors, *Coord. Chem. Rev.*, *250*, 2939-51.

[54] Davis, A. P. (1993). Cholaphanes et al.; steroids as structural components in molecular engineering, *Chem. Soc. Rev.*, *22*, 243-53.

[55] Carey, M. C. (1985). Physical-chemical properties of bile acids and their salts, *Sterols and bile Acids.H. Danielsson, J. Sjövall. Elsevier Sci. Publ. 5.*

[56] Coello, A., Meijide, F., Rodríguez Núñez, E. & Vázquez Tato, J. (1996). Aggregation Behavior of Bile Salts in Aqueous Solution, *J. Pharm. Sci.*, *85*, 9-15.

[57] Betebenner, D. A., Carney, P. L., Zimmer, A. M., Kazikiewicz, J. M., Brucher, E., Sherry, A. D. & Johnson, D. K. (1991). Hepatobiliary delivery of polyaminopolycarboxylate chelates: synthesis and characterization of a cholic acid conjugate of EDTA and biodistribution and imaging studies with its indium-111 chelate, *Bioconjugate Chem.*, *2*, 117-23.

[58] Subramanian, K. M. & Wolf, W. (1990). A new radiochemical method to determine the stability constants of metal chelates attached to a protein, *J. Nucl. Med.*, *31*, 480-8.

[59] Armstrong, M. J. & Carey, M. C. (1982). The hydrophobic-hydrophilic balance of bile salts. Inverse correlation between reverse-phase high performance liquid chromatographic mobilities and micellar cholesterol-solubilizing capacities, *J. Lipid Res.*, *23*, 70-80.

[60] Coello, A., Meijide, F., Rodríguez Núñez, E. & Vázquez Tato, J. (1993). Aggregation behavior of sodium cholate in aqueous solution, *J. Phys. Chem.*, *97*, 10186-91.

[61] Enhsen, A., Kramer, W. & Wess, G. (1998). Bile acids in drug discovery, *Drug Discovery Today*, *3*, 409-18.

[62] Herraez, E., Macias, R. I. R., Vázquez Tato, J., Vicens, M., Monte, M. J. & Marin, J. J. G. (2009). In vitro inhibition of OATP-mediated uptake of phalloidin using bile acid derivatives, *Toxicol. Appl. Pharm.*, *239*, 13-20.

[63] Herraez, E., Macias, R. R., Vázquez Tato, J., Vicens, M., Monte, M. J. & Marin, J. J. G. (2009). Protective effect of bile acid derivatives in phalloidin-induced rat liver toxicity, *Toxicology Appl. Pharm.*, *239*, 21-28.

[64] Alvarez Alcalde, M. (2007). Estructuras supramoleculares generadas por ácidos biliares y ciclodextrinas, *Estructuras supramoleculares generadas por ácidos biliares y ciclodextrinas, Tesis Doctoral, USC.*

[65] Gouin, S. & Zhu, X. X. (1998). Fluorescence and NMR Studies of the Effect of a Bile Acid Dimer on the Micellization of Bile Salts, *Langmuir*, *14*, 4025-9.

[66] Alvarez Alcalde, M., Jover, A., Meijide, F., Galantini, L., Pavel, N. V., Antelo, A. & Vázquez Tato, J. (2008). Synthesis and Characterization of a New Gemini Surfactant Derived from 3a ,12a -Dihydroxy-5b -cholan-24-amine (Steroid Residue) and Ethylenediamintetraacetic Acid (Spacer), *Langmuir*, *24*, 6060-6.

[67] Álvarez Alcalde, M., Jover, A., Meijide, F., Galantini, L., Pavel, N. V., Antelo, Á. & Vázquez Tato, J. (2009). Aggregation Behavior of Tetracarboxylic Surfactants Derived from Cholic and Deoxycholic Acids and Ethylenediaminetetraacetic Acid, *Langmuir*, *25*, 9037-44.

[68] Fini, A., Fazio, G., Roda, A., Bellini, A. M., Mencini, E. & Guarneri, M. (1992). Basic cholane derivatives. XI: comparison between acid and basic derivatives, *J. Pharm. Sci.*, *81*, 726-30.

[69] Meijide, F., Antelo, A., Alvarez, M., Soto, V. H., Trillo, J. V., Jover, A. & Vázquez Tato, J. (2011). Spontaneous formation in solid state of carbamate derivatives of bile acids, *Crystal Growth Des.*, *11*, 356-61.

[70] Anelli, P. L., Lattuada, L. & Uggeri, F. (1998). One-pot Mitsunobu-Staudinger preparation of 3-aminocholan-24-oic acid esters from 3-hydroxycholan-24-oic acid esters, *Synth. Commun.*, *28*, 109-17.

[71] Kawamura, H., Murata, Y., Yamaguchi, T., Igimi, H., Tanaka, M., Sugihara, G. & Kratohvil, J. P. (1989). Spin-label studies of bile salt micelles, *J. Phys. Chem.*, *93*, 3321-6.

[72] Soto Tellini, V. H. J., Aida, Carrazana Garcia, Jorge, Galantini, Luciano, Meijide, Francisco, Vázquez Tato & José. (2006). Thermodynamics of Formation of Host-Guest Supramolecular Polymers, *J. Am. Chem. Soc.* *128*, 5728-34.

[73] Scholtissek, C., Quack, G., Klenk, H. D. & Webster, R. G. (1998). How to overcome resistance of influenza A viruses against adamantane derivatives, *Antiviral Res.*, *37*, 83-95.

[74] Vashkevich, E. V., Yurashevich, N. Y., Kozlov, N. G., Potkin, V. I. & Potkina, T. N. (2001). Synthesis of surfactants derived from adamantane, *Russ. J. Appl. Chem. (Translation of Zhurnal Prikladnoi Khimii)*, *74*, 1892-8.

[75] Trillo Novo, J. V. (2013). unpublished work.

[76] Soto, V. H., Jover, A., Galantini, L., Meijide, F. & Vazquez-Tato, J. (2005). Synthesis and complexation of polytopic adamantane-based probes, *Proceedings of ECSOC-9, International Electronic Conference on Synthetic Organic Chemistry, 9th,*, Nov. 1-30. F004.

[77] Szejtli, J. (1988). Kluwer Academic Publishers, Dordrecht, The Nederlands.

[78] Bender, M. L. & Komiyama, M., (Eds.) (1978). *Cyclodextrin chemistry*, Springer-Verlag, Berlin.

[79] Saenger, W., Jacob, J., Gessler, K., Steiner, T., Hoffmann, D., Sanbe, H., Koizumi, K., Smith, S. M. & Takaha, T. (1998). Structures of the

Common Cyclodextrins and Their Larger Analogs -Beyond the Doughnut, *Chem. Rev.*, *98*, 1787-802.

[80] Carrazana, J., Jover, A., Meijide, F., Soto, V. H. & Vázquez Tato, J. (2005) Complexation of Adamantyl Compounds by b-Cyclodextrin and Monoaminoderivatives, *J. Phys. Chem. B*, *109*, 9719-26.

[81] Ciferri, A. e. (2005). Supramolecular Polymers, *Supramolecular Polymers, 2nd edition Taylor Francis, New York.*

[82] Capretta, A., Maharajh, R. B. & Bell, R. A. (1995). Synthesis and characterization of cyclomaltoheptaose-based metal chelants as probes for intestinal permeability, *Carbohydr. Res.*, *267*, 49-63.

[83] Yan, J.-M., Atsumi, M., Yuan, D.-Q. & Fujita, K. (2002). (Ethylenediaminetetraacetic acid)cerium(IV) [CeIV(EDTA)] complexes with dual hydrophobic binding sites as highly efficient catalysts for the hydrolysis of phosphodiesters, *Helv. Chim. Acta*, *85*, 1496-504.

[84] Michels, J. J., Huskens, J. & Reinhoudt, D. N. (2002). Noncovalent Binding of Sensitizers for Lanthanide(III) Luminescence in an EDTA-bis(b-cyclodextrin) Ligand, *J. Am. Chem. Soc.*, *124*, 2056-64.

[85] Mulder, A., Huskens, J. & Reinhoudt, D. N. (2005). Complexation of charged porphyrins by charged and metal-chelated EDTA-tethered b -cyclodextrin dimers: a thermodynamic study on the influence of tether charge and flexibility on binding affinity, *Eur. J. Org. Chem*, 838-46.

[86] Wathier, M. & Grinstaff, M. W. (2008). Synthesis and Properties of Supramolecular Ionic Networks, *J. Am. Chem. Soc.*, *130*, 9648-9.

[87] Antelo, A., Alvarez, M., Jover, A., Meijide, F., Galantini, L. & Vázquez Tato, J. (2009). Complexation of adamantyl derivatives by a β-cyclodextrin dimer, *International Electronic Conference on Synthetic Organic Chemistry*, 13th, Nov. 1-30, 2009, ante3/1-ante3/7.

[88] Antelo, A., Alvarez Alcalde, M., Jover, A., Meijide, F., Galantini, L. & Vázquez Tato, J. (2008). Supramolecular polymer formed by ditopic cyclodextrin and ditopic adamantane derivatives, 12th ECSOC November 1-30

[89] Hocquelet, C., Jankowski, C. K., Pelletier, A. L., Tabet, J.-C., Lamouroux, C. & Berthault, P. (2011). Synthesis and inclusion properties study of some mono 6-amino β-cyclodextrin dimers bridged by N,N-succinyldiamide linkers, *J. Incl. Phenom. Macrocycl. Chem.*, *69*, 75-84.

[90] Mulder, A., Auletta, T., Sartori, A., Del Ciotto, S., Casnati, A., Ungaro, R., Huskens, J. & Reinhoudt, D. N. (2004). Divalent Binding of a Bis(adamantyl)-Functionalized Calix[4]arene to b -cyclodextrin-based

Hosts: An Experimental and Theoretical Study on Multivalent Binding in Solution and at Self-Assembled Monolayers, *J. Am. Chem. Soc., 126*, 6627-36.

[91] Alvarez Parrilla, E., Ramos Cabrer, P., Al-Soufi, W., Meijide, F., Rodríguez Núñez, E. & Vázquez Tato, J. (2000). Dendritic growth of a supramolecular complex, *Angew. Chem. Int., Ed., 39*, 2856-8.

[92] Soto Tellini, V. H., Jover, A., Galantini, L., Meijide, F. & Vázquez Tato, J. (2004). Crystal structure of the supramolecular linear polymer formed by the self-assembly of mono-6-deoxy-6-adamantylamide-β-cyclodextrin, *Acta Crystal., B60*, 204-10.

[93] Antelo, A., Jover, A., Galantini, L., Meijide, F., Alvarez Alcalde, M., Pavel, N. V. & Vázquez Tato, J. (2011). Formation of host-guest and sandwich complexes by a β-cyclodextrin derivative, *J. Inclusion Phenom. Macrocycl. Chem., 69*, 245-53.

[94] Ottenbrite, R. M. & Hu, B. (2000). Synthesis and characterization of PEG-EDTA bidentates as oral drug systems, *Polymer Preprints, 41*, 6-7.

[95] Alvarez Alcalde, M. A., Alvaro, Jover, Aida, Meijide, Francisco, Gancedo, Cristina, Galantini, Luciano & Vázquez Tato' José. (2009). Solubilization of Cholesterol in Aqueous Solution by two β-cyclodextrin dimers and a negatively charged β-cyclodextrin derivative, *J. Inclusion Phenom. Macrocycl. Chem., 63*, 309-17.

[96] Jiang, H., Zhang, S. & Shi, Q. (2011). Removal of cholesterol by β-cyclodextrin, *Asian J. Chem., 23*, 3783-6.

[97] Kwak, H. S., Jung, C. S., Shim, S. Y. & Ahn, J. (2002). Removal of Cholesterol from Cheddar Cheese by β-cyclodextrin, *J. Agric. Food Chem., 50*, 7293-8.

[98] Maffeo, D., Lampropoulou, M., Fardis, M., Lazarou, Y. G., Mavridis, I. M., Mavridou, D. A. I., Urso, E., Pratsinis, H., Kletsas, D. & Yannakopoulou, K. (2010). Novel polycarboxylated EDTA-type cyclodextrins as ligands for lanthanide binding: study of their luminescence, relaxivity properties of Gd(III) complexes, and PM3 theoretical calculations, *Org. Biomol. Chem., 8*, 1910-21.

[99] Dazzi, J. (1964). Bis(2,6-dioxopiperazinyl)alkane, *Patent*, DE 1180372 19641029 US3196153.

[100] Herbst, S. & Eckert, H. W. (1972). Textile softening quaternized dioxopiperazines, *Patent*, DE 2064405 A 19720713.

[101] Woo, E. P. (1985). Polyamide from bis-carboxy containing 3,5,3',5'-tetraoxo-1,2-dipiperazine compound and diamine, *Patent*, US 4536564 A 19850820.

[102] Houghton, R. P. & Williams, E. (1982). Synthesis of bis(imides) and bis(half amides) of N,N'-ethylenebis(iminodiacetic acid), *J. Chem. Soc. Perkin Trans*, *1*, 2693-6.

[103] Frankus, E., Hellmann, K. & Mueckter, H. (1976). 2,6-Dioxopiperazine derivatives, *Patent*. DE 2511891 A1 19761007.

[104] Takeshita, T., Maeda, S., Awata, K., and Imayoshi, M. (1981) Studies on the utilization of metal chelate. XII. Preparation of acrylic polymers containing chelating ligands as pendent groups, *Yukagaku*, *30*, 447-9.

[105] Hwang, D.-C. & Damodaran, S. (1996). Chemical Modification Strategies for Synthesis of Protein-Based Hydrogel, *J. Agric. Food Chem.*, *44*, 751-8.

[106] Maeda, S., Tsurusaki, Y., Tachiyama, Y., Naka, K., Ohki, A., Ohgushi, T. & Takeshita, T. (1994). Synthesis and properties of superconductors prepared from ethylenediaminetetraacetic acid (EDTA)-ethylenediamine (ED) polyamide YBC chelate, *J. Polym. Sci. A*, *32*, 1729-38.

[107] Naka, K., Tanaka, Y., Yamasaki, K., Ohki, A., Chujo, Y. & Maeda, S. (2001). Synthesis of soluble complexan polymers in organic solvents for using as a polymer-chelate precursor to YBa2Cu3O7-x thin films, *Bull. Chem. Soc. Jpn*, *74*, 571-7.

[108] Imai, Y. (1970). Poly-2,6-piperazinedione. New class of polymer derived from N,N,N',N'-ethylenediaminetetraacetic acid dianhydride and diamines, *Makromol. Chem.*, *138*, 293-7.

[109] Di Noto, V., Munchow, V., Vittadello, M., Collet, J. C. & Lavina, S. (2002). Synthesis, characterization and conductivity studies of Li and Mg polymer electrolytes based on esters of ethylenediaminetetraacetic acid and PEG 400, *Solid State Ionics*, *147*, 397-402.

[110] Di Noto, V. (2002). Electrical Spectroscopy Studies of Lithium and Magnesium Polymer Electrolytes Based on PEG 400, *J. Phys. Chem. B* *106*, 11139-54.

[111] Di Noto, V., Munchow, V., Vittadello, M., Collet, J. C. & Lavina, S. (2002). Synthesis and characterization of lithium and magnesium complexes based on [EDTA][PEG400]2 and [EDTA]3[PEG400]7, Macromol. Chem. Phys., 203, 1211-27.

[112] Kahana, N., Arad-Yellin, R. & Warshawsky, A. (1994). A Conceptual Approach to the Synthesis of Bifunctional EDTA Analogs: EDTA-Extended Polyamides, *J. Org. Chem.*, *59*, 4832-7.

[113] Buschhaus, B., Hampel, F., Grimme, S. & Hirsch, A. (2005). Metal-induced chiral folding of depsipeptide dendrimers, *Chem. Eur. J.*, *11*, 3530-40.

In: EDTA
Editor: Andris Molnar
ISBN: 978-1-62808-146-6

Chapter 2

STRUCTURAL AND ELECTRICAL PROPERTIES OF $CaCu_3Ti_4O_{12}$ OBTAINED FROM DIFFERENT EDTA-GEL DERIVED POWDERS

Filipe Amaral[1,2], Manuel Valente[2], Manuel Pedro Graça[2] and Luís Cadillon Costa[2]

[1]Polytechnic Institute of Coimbra, Coimbra, Portugal
[2]I3N and Physics Department, University of Aveiro, Aveiro, Portugal

ABSTRACT

High dielectric permittivity has been found on the oxide $CaCu_3Ti_4O_{12}$ (CCTO), in the order of 10^5 at 1 kHz and room temperature, being nearly constant to temperatures until 300 °C. These characteristics lead this material to be useful in a high number of technological applications, including the miniaturization. The physical explanation for the high dielectric permittivity behaviour is actually an open question. Some researchers suggest that this behaviour is intrinsic, while others attribute to extrinsic effects. Within this work we report the synthesis and characterization of CCTO powders prepared by two different wet-chemical methods: ethylenediaminetetraacetic acid-gel method (EDTA) and modified Pechini (MP) process. In both cases, almost single-phase CCTO fine powders were obtained from the calcination of the gel precursors at lower temperatures and shorter times than those reported for the conventional solid state reaction process. Also thin films were produced using spin-coating technique, depositing the obtained gels on Si(111) substrates. The structure of the calcinated powders, sintered pellets and thin films were characterized by X-ray diffraction and scanning electron microscopy. Impedance spectroscopy measurements were performed on the sintered pellets and films, in the frequency range from 40 Hz to 110 MHz, at room temperature. The results are discussed and compared. High frequency relaxation data was fitted with Cole–Cole model and the calculated relaxation parameters were correlated with morphological and structural properties.

1. INTRODUCTION

1.1. High Dielectric Constant Materials

The global market for capacitors represented in 1993 an amount close to US$ 12.3 billion [39]. With the increasing of electronics market and associated demand for new technologies, the development of new dielectric materials and production methods has shown a significant growth during the last decades [41, 27]. It was exactly the advent of information and communication technologies that triggered a huge demand for smaller and portable devices, with higher autonomy and better performance, which would need the development of new high dielectric constant (ε') materials [38]. These materials have multiple applications, including gate dielectrics, high charge-storage capacitor and electroactive materials [32]. Within the multiple capacitor classes, ceramic capacitors represent the largest sector, regarding global sales, followed by aluminum, paper and film, and tantalum [52].

The simplest capacitor can be modeled as two parallel conductive plates separated by an insulating (dielectric) material (figure 1). When a voltage V is applied between the parallel plates, a static electric field is developed across the dielectric, causing the accumulation of electrical charges (Q) at the conductive plates, so that $Q = CV$, where C is known as the capacitance. In the case of the parallel plate capacitor, the capacitance can be calculated using the following expression:

$$C = \frac{\varepsilon' \varepsilon_0 A}{d} \tag{1}$$

where ε' is the dielectric constant of the insulator material, ε_0 is the permittivity of vacuum, A is the surface area of the conductive plates and d is the thickness of the dielectric.

Another important characteristic of capacitors is the dielectric loss (tan δ), defined as the ratio between the imaginary and real parts of the complex permittivity ($\varepsilon^* = \varepsilon' - j\varepsilon''$). It refers to the power loss resulting from the phase difference between the applied a.c. voltage and the current. In this case, the lower is tan δ the better will be the dielectric material.

So far, among ceramic capacitors, ferroelectric materials with perovskite type crystalline structure, like $BaTiO_3$, proved to have excellent properties for applications in high dielectric constant capacitors [53, 42, 22]. Like all other ferroelectrics, when $BaTiO_3$ ceramics are heated above the Curie point ($T \approx$ 130 °C), the crystalline structure goes through a phase transition from tetragonal to cubic, without any spontaneous polarization, reaching a paraelectric state. Below that temperature, a phase transition from cubic to tetragonal occurs because of the stretching of c axis in the axial direction. The other axes shrink slightly, forming the spontaneous polarization. These changes produce additional stress over the grains that generate small polarized domains (ferroelectric phase). Applying an appropriate electric field it will be possible to reverse the spontaneous polarization, increasing the dielectric constant of the ceramic. Although these ferroelectric ceramics can present high dielectric constant values, the existence of phase transitions can be a problem when capacitors are used in a large range of temperatures, attending to the abrupt variations of ε' around phase transition points.

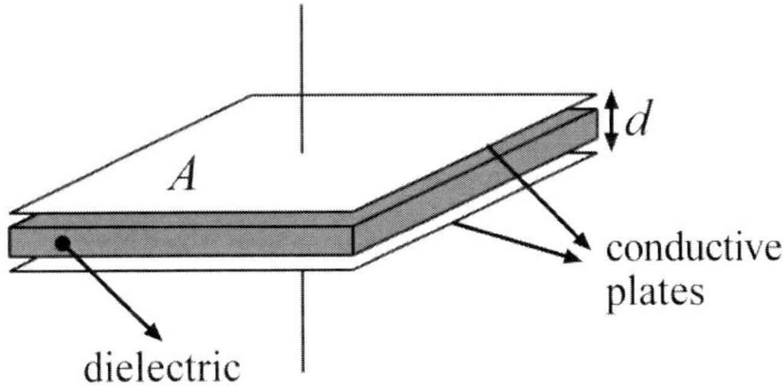

Figure 1. Parallel plate capacitor model.

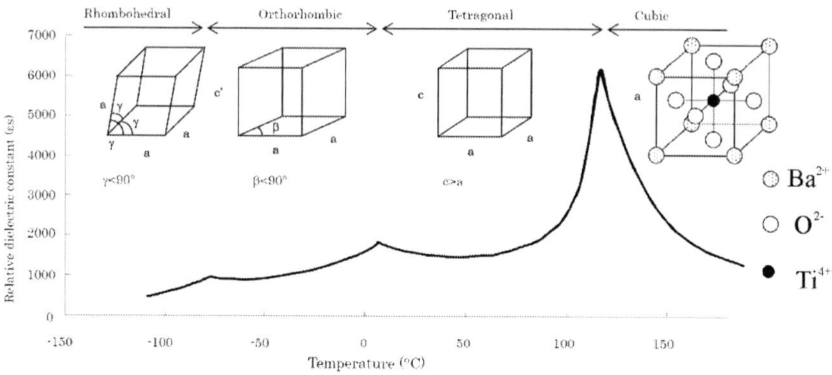

Figure 2. Dependence of crystalline structure of $BaTiO_3$ and relative dielectric constant on temperature.

$CaCu_3Ti_4O_{12}$ (CCTO), recently rediscovered by Subramanian et al. [48], is a non-ferroelectric material that is being pointed as very attractive from both a technological and a scientific standpoint. CCTO presents a perovskite structure (figure 3), with the particularity of having very high ε' values and good stability over a wide temperature range (100 - 400 K) and frequency (100 Hz - 1 MHz), which represents a major advantage when compared to ferroelectric ceramics.

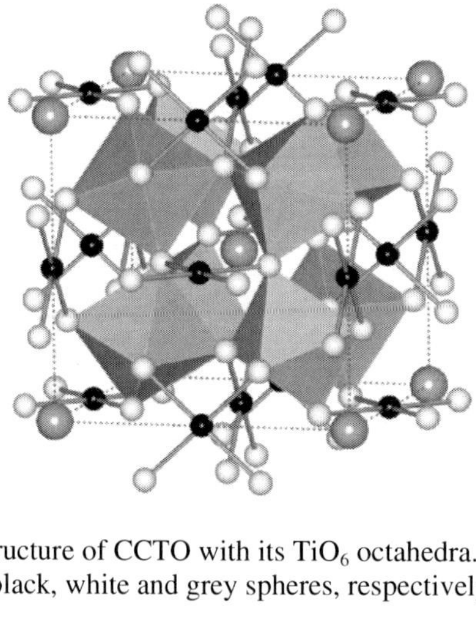

Figure 3. Crystalline structure of CCTO with its TiO_6 octahedra. The Cu, O and Ca atoms are shown with black, white and grey spheres, respectively [49].

The high dielectric constant (HDC) was observed in ceramics, single crystals and films of CCTO. Below 100 K, the dielectric constant has a step decrease of 3 orders of magnitude and a peak in the dielectric loss (tan δ), which is frequency dependent. CCTO shows strong nonlinear current-voltage behaviour. No change is detected in the crystal structure by Neutron Powder Diffraction between 35 and 1273 K. However, no explanations are offered as to why, among the many titanates with perovskite structure, CCTO shows a HDC regardless of how it is prepared and the type of electrode used. The main drawback of CCTO, for practical applications, is the high tan δ (tan $\delta \approx 0.1$ at 1 kHz and 300 K).

The CCTO properties justify, since the first report [48], the great interest in this material. For the dielectric properties, the Internal Barrier Layer Capacitor (IBLC) model, where insulating surfaces coat semiconducting grains, describes well the dielectric behaviour of ceramic materials, but cannot be applied to single crystals [2, 3, 4]. The dielectric constant of single crystals is higher than the dielectric constant of polycrystals.

The CCTO microstructure and dielectric properties are strongly dependent of the dopants, sintering parameters and synthesis method.

The dopants of CCTO, which act as acceptors, have a critical impact on the disappearance of the IBLC [29, 10, 16]. As an example, ε' can be increased to $1.5\text{x}10^5$ with 5% of Co, and the doping with ZrO_2 reduces the dielectric loss, maintaining the value of the dielectric constant. With a suitable choice of dopant, the dielectric values of the pure CCTO can be improved, making the doped CCTO more interesting for applications in the industry [44], thanks to the increasing of the grain size and/or to the depressing of the barrier width [33].

Several studies [47, 44] point to the influence of the sintering parameters over the dielectric, electrical and microstructural properties of CCTO. Most of the studies refer that higher sintering temperatures (until the fusion point) and longer sintering periods, contribute to a higher ε'. The heat-treatment, in controlled atmosphere, can modify the dielectric response of CCTO [14, 15]. For example, the heat-treatment in oxygen-rich atmosphere improves the non-ohmic behaviour and annealing at oxygen poor atmosphere decreases the non-ohmic properties of CCTO.

The study of CCTO properties up to 2013 produced numerous articles, mainly during the last 12 years, most of them evolving the analysis of ceramic samples produced by solid state method. The use of alternative methods, like wet-chemistry methods, for the production of CCTO powders and films still needs further research, particularly on the optimization of the synthesis parameters, so that critical points related to microstructural control can be solved.

Within this chapter we report the synthesis and characterization of CCTO powders prepared by two different wet-chemical methods: ethylenediaminetetraacetic acid-gel method (EDTA) and modified Pechini (MP) process. In both cases, almost single-phase CCTO fine powders were obtained from the calcination of the gel precursors at lower temperatures and shorter times than those reported for the conventional solid state reaction process. The structure of the calcined powders and sintered pellets were characterized by X-ray diffraction (XRD) and scanning electron microscopy (SEM). Impedance spectroscopy measurements were performed on the sintered pellets, in the frequency range from 40 Hz to 110 MHz, at several temperatures. The results are discussed and compared. High frequency relaxation data was fitted with Cole Cole model and the calculated relaxation parameters were correlated with morphological and structural properties.

1.2. Microstructural-Electrical Model of Polycrystalline Ceramic

Polycrystalline ceramic materials can be placed into a larger group of materials generically called non-homogeneous media. Although the polycrystalline ceramic could be formed by monophasic grains, the composition and properties of the surface of the samples and grain boundaries are, normally, distinct from those of the grains. This kind of microstructure results in the existence of different electrical conductivities between those structures, which originates interfacial polarizations and, consequently, the arising of space charge regions.

Several models, representing the electrical behavior of these structures, have been proposed [36, 7]. Among those, the pioneering study of Bauerle [7] constitutes the fundamental basis on correlating microstructure and electrical properties of polycrystalline materials. Bauerle presented an equivalent electric circuit, representing the electrical behavior of the polarizations of the grains, grain boundaries and sample surface-electrode interface, consisting of the series connection of three RC parallel elements, each one representing the electrical response of grain (R_g, C_g), grain boundaries (R_{gb}, C_{gb}) and sample-electrode interface (R_e, C_e), as it is illustrated at figure 4. Microstructural models that predict an electrical behavior like that presented by Bauerle's Circuit Equivalent can be grouped into two classes: Layer Models and Effective Medium Models. Due to its simplicity, the Brick Layer Model, proposed by Beeckman and Heyne [8] and later developed by Dijk and Burgraaf [13], is one of the most useful tools in the characterization of polycrystalline samples.

The Brick Layer Model (BLM) views the microstructure of polycrystalline ceramic as a 3D lattice formed by cubic shaped grains (bricks), with an edge size d_g, distributed through a simple cubic lattice, separated by grain boundaries with a thickness d_{gb}. Each unit cell can be decomposed in two distinct parts, as illustrates figure 5: one constituted by the cubic grains and the grain boundary top faces, and another one corresponding to the lateral coating of the grain with the grain boundary material. Whenever the resistivity of the grain boundaries is much higher than that of the grains, like in CCTO, it is possible to neglect the current that passes through the lateral walls of the unit cell, that is, we only consider the current passing through the series track, as illustrated at figure 5.

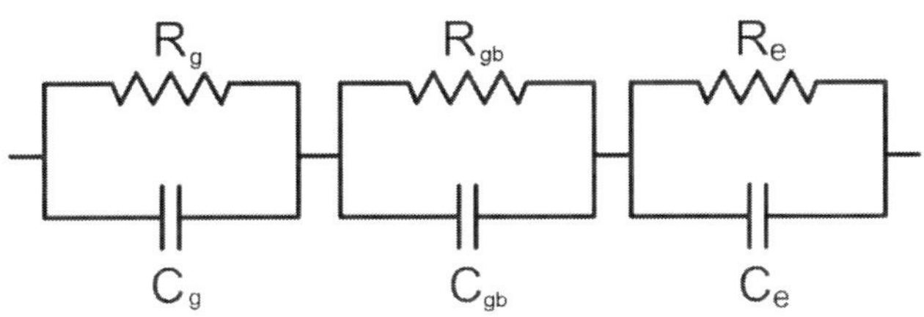

Figure 4. Circuit equivalent for a polycrystalline ceramic according to Baurle [7].

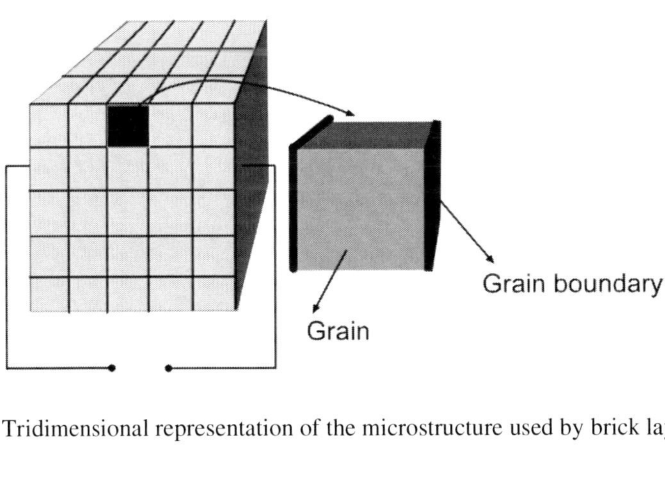

Figure 5. Tridimensional representation of the microstructure used by brick layer model.

In these conditions, it is possible to predict the resistance of the grain boundaries (R_{gb}), in agreement with the following expression:

$$R_{gb} = \frac{\rho_{gb} d_{gb} N}{A}, \tag{2}$$

where ρ_{gb} is the resistivity of the grain boundaries, A is the electrode surface area and N the number of grain boundaries found along a sample of thickness L ($N = \frac{L}{(d_g + d_{gb})} \approx \frac{L}{d_g}$). Using this expression and equation (2) it is found that:

$$R_{gb} = \frac{L \rho_{gb} d_{gb}}{d_g A}. \tag{3}$$

Considering that macroscopic conductivity (σ_{gb}^{mac}) is given by $\frac{1}{R_{gb}} \frac{L}{A}$, and combining it with equation (3), it can be found the following relation between microscopic conductivity ($\sigma_{gb}^{mic} = \frac{1}{\rho_{gb}}$) and σ_{gb}^{mac}:

$$\sigma_{gb}^{mac} = \sigma_{gb}^{mic} \frac{d_g}{d_{gb}}. \tag{4}$$

The expression (4) shows that if the intrinsic properties of the grain boundaries remain unchanged, that is $\frac{\sigma_{gb}^{mic}}{d_{gb}} = constant$, the macroscopic conductivity of the polycrystalline ceramic increases linearly with the grain size, d_g. It must be mentioned that d_g represents the average grain size.

It is also possible to set up a relation between the electrical capacitance of each grain boundary (C_i in equation (5)), with the electrical capacitance of all boundaries (C_{gb}), considering a series association of those structures.

$$C_i = \varepsilon_0 \varepsilon_{r_{gb}} \frac{A}{d_{gb}}. \tag{5}$$

Considering that the number of grain boundaries is $N \approx \frac{L}{d_g}$, the following equation for this capacitors series association can be considered:

$$\frac{1}{C_{gb}} = \sum_{j=1}^{N} \frac{1}{C_j} \Leftrightarrow C_{gb} = \frac{d_g C_i}{L} \quad (6)$$

Replacing the equation (5) in (6) and considering $C_0 = \varepsilon_0 \frac{A}{L}$, we obtain:

$$\frac{C_{gb}}{C_0} = \varepsilon_{r_{gb}} \frac{d_g}{d_{gb}}. \quad (7)$$

From this equation it can be found that if the intrinsic properties of the grain boundaries remain invariable, that is if $\frac{\varepsilon_{r_{gb}}}{d_{gb}} = constant$, the macroscopic capacitance (C_{gb}) of a polycrystalline ceramic should linearly increase with the increase of the grain size (d_g).

Although the utility of BLM, the existence of cubic grains in a real ceramic microstructure is rather improbable. Several deviations from BLM were predicted by Fleig and Maier [17, 18, 19, 20], namely:

a) Deviations from the cubic grain shape;
b) Imperfect contacts between the grains;
c) Grain boundary properties varying from boundary to boundary.

Fleig and Maier used finite element method to analyze the effects of different microstructural patterns on the electric flux and current distributions generated in the polycrystalline microstructure. Among the multiple deviations predicted by Fleig and Maier simulations, we emphasize the fact that non homogenous grain size distributions, as bimodal distributions, contribute to a decrease of the expected sample resistance because of the "detour-effect". This effect is related with the deflection of the flux lines when tracking lower impedance routes, tending to pass through grains of larger dimensions, reducing by this way the number of crossed grain boundaries.

1.3. Dielectric Measurements

Because of the importance of this technique for the dielectric characterization of polycrystalline ceramics, we will briefly resume the fundamentals of impedance spectroscopy. This technique analyses the dielectric response obtained from the different entities that constitute the microstructure of a system, based on the fact that each entity should present a particular relaxation time, which allow differentiating each one of the dielectric responses.

When a sinusoidal voltage $v(t) = V_m sen(\omega t)$, with frequency $f = \frac{\omega}{2\pi}$, is applied to a material, it is generated an electric current $i(t) = I_m sen(\omega t + \phi)$ that goes through the material with a certain offset ϕ relatively to v(t) .

According to Ohm's law, the impedance of a system (Z*(ω)) may be expressed as:

$$Z^*(\omega) = \frac{v^*(t)}{i^*(t)} = \frac{V_m sen(\omega t)}{I_m sen(\omega t + \phi)} = Ze^{-j\phi} \quad (8)$$

so that
$$Z^*(\omega) = z\cos\phi - jz sen\phi = Z_R - jZ_I \quad (9)$$

The amplitude of this complex quantity will be:

$$|Z^*| = \sqrt{Z_R^2 + Z_I^2} \text{ and } \phi = \tan^{-1}\left(\frac{Z_I}{Z_R}\right) \quad (10)$$

We should remember that for a pure resistance (with a resistance *R*) and a pure capacitor (with a capacitance *C*), corresponds an impedance equal to $Z^*(\omega) = R$, $Z^*(\omega) = -j(\omega C)^{-1}$, respectively.

Experimental data can be presented in the complex plane using different formalisms, as for example: the complex impedance (*Z**), the complex admittance (*Y**), the complex permittivity (ε^*) and the modulus function (*M**). The mathematical relation between each one is shown in table 1.

Table 1. Relation between *M, *Z**, *Y** and ε^*. $\gamma = j\omega C_0$, where C_0 is the capacitance of the empty cell**

	M^*	Z^*	Y^*	ε^*
M^*	-	γZ^*	γY^{*-1}	ε^{*-1}
Z^*	$\gamma^{-1} M^*$	-	Y^{*-1}	$\gamma^{-1}\varepsilon^{*-1}$
Y^*	γM^{*-1}	Z^{*-1}	-	$\gamma\varepsilon^*$
ε^*	M^{*-1}	$\gamma^{-1}Z^{*-1}$	$\gamma^{-1}Y^*$	-

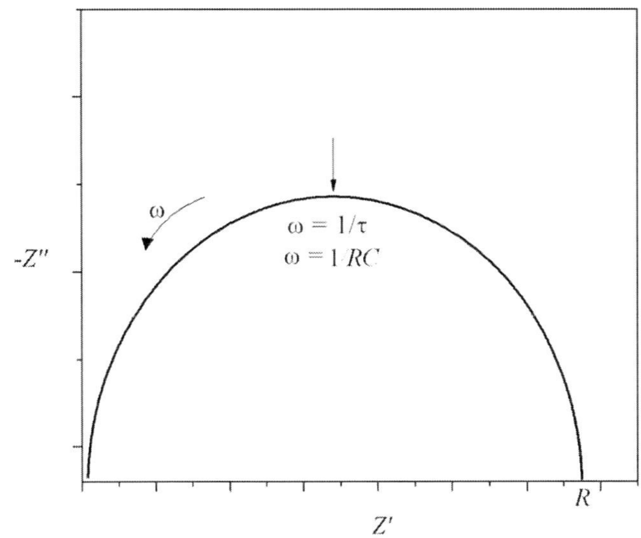

Figure 6. Nyquist plot of complex impedance (Z*) obtained for the parallel *RC* circuit.

From the analysis of the graphs obtained on the complex plane it is possible to isolate the dielectric/electric response of the different entities that form the studied material in the used frequency domain.

Generally, using the right formalism and a large frequency and temperature ranges, becomes possible to isolate the response of those several entities (bulk and the different interfaces).

The understanding of the dielectric relaxation phenomena is easier if we use a simple circuit formed by the parallel association of a resistance (*R*) and a capacitor (*C*) to study the dielectric response of that system when submitted to a sinusoidal voltage.

From the representation of Z'' vs. Z', also called the Nyquist plot of Z^*, it results a semicircle whose centre is placed at the real axis. The intersection of the semicircle with the real axis at the lower frequency region corresponds to the value of *R*. The frequency at which a maximum value of Z'' is achieved allows the achievement of the capacitance (*C*), once $2\pi f_{max} = \frac{1}{\tau} = \frac{1}{RC}$.

Similarly, also the dielectric response of the polycrystalline ceramics, when using the Nyquist representation of Z^*, appears in the form of semicircles, where each semicircle represents the dielectric behavior of a different phenomena/structure.

However, it is not always easy to distinguish the totality of the evolved phenomena. If the relaxation frequencies are similar there will be an overlapping of the semicircles. When this occurs, it becomes necessary to establish measurements in a larger temperature range so that the isolation of the several relaxations can succeed. Obviously, this procedure assumes that the evolved structures will respond differently to variations of temperature.

The simplest representation for a polycrystalline system corresponds to the series combination of *N* sub-circuits, each of them including a parallel association of a resistance with a capacitor (figure 4). If each of this sub-circuits has a relaxation frequency far different from the other, than the electric response of the system will correspond to a set of *N* semicircles, separated from each other, as illustrates figure 7.

In the case of polycrystalline ceramics it is common to use three *RC* sub-circuits, where the first semicircle, appearing in the high frequency region (at the left), corresponds to the dielectric response of the bulk (i.e. the grains), the intermediate semicircle is associated to the response of grain boundaries and, finally, the last is associated to the response of the interface bulk/electrode.

Usually, the observed arcs are not centered in the *Z´* axis, confirming that a single relaxation time, described by the Debye equation, cannot be used to explain that relaxation phenomenon. Because the majority of our samples have this kind of dielectric behavior, we have fitted the data with the Cole-Cole function (model) [11]:

$$Z^{*}(\omega) = \frac{R}{1 + \left(i\omega\tau_{cc}\right)^{1-\alpha}} \tag{11}$$

In this model, which is an empirical modification of the Debye equation, *R* is the resistance of the studied structure, τ_{cc} the relaxation time ($\tau_{cc}=RC$) and α a parameter between 0 and 1 that reflects the dipole interaction. Figure 8 shows the graphical representation of the Cole-Cole function. For determining the parameters τ_{cc} and α, we first calculate the approximate values from the asymptotic part of the data, and then use it as starting values in a non-linear curve fitting algorithm [12]. When α is close to 0, the material is more homogeneous, the dipole interaction is low, and the semicircle is practically centered in the *Z´* axis. The angle of depression can be defined as [26]:

$$\Phi = \alpha \frac{\pi}{2} \tag{12}$$

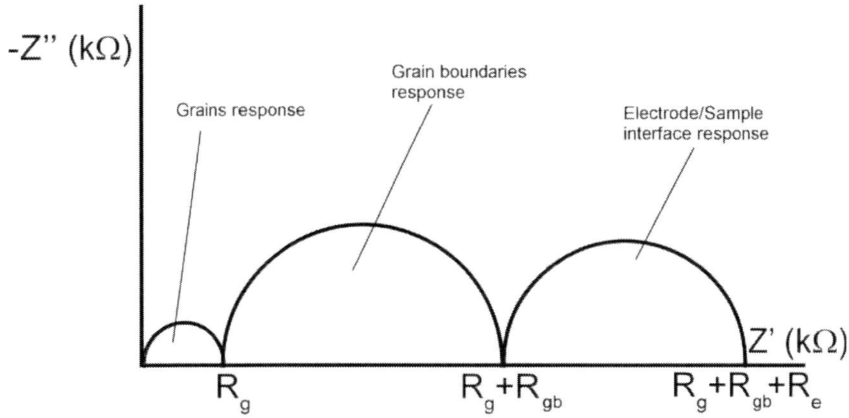

Figure 7. Nyquist representation of the impedance response of a typical polycrystalline ceramics.

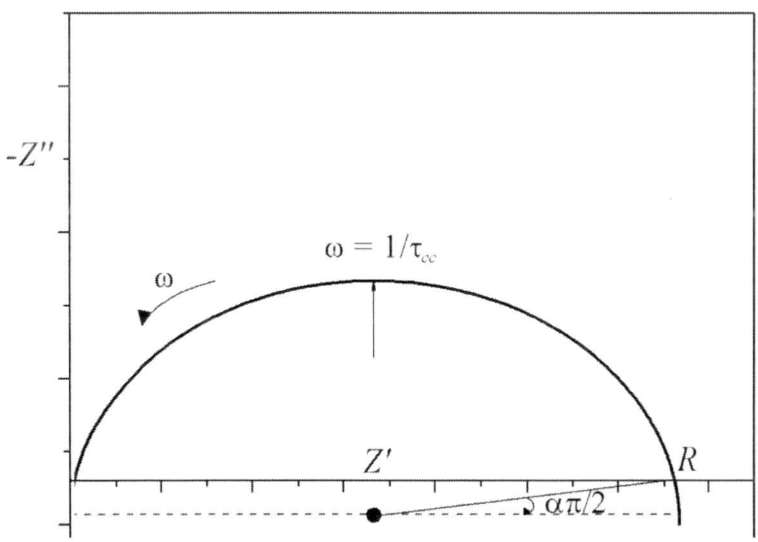

Figure 8. Nyquist representation of complex permittivity (ε^*) for a Cole-Cole type relaxation.

1.4. Wet-Chemical Methods

The use of wet-chemical methods allows the production of ceramic powders with high purity level and homogeneity, with precise stoichiometry and, generally, requiring shorter reaction times and lower calcination temperatures to achieve the wanted phase [37, 9].

Among the several alternative ways to the preparation of oxides by sol-gel we can include those using citric acid (2-hydroxypropane-1,2,3-tricarboxylic acid) as chelating agent. This sol-gel method is usually known as Pechini method [40]. In this case, a solution with ethylene glycol, citric acid and metallic ions form, once polymerized, a polyester type resin. This technique allows that a large sort of metallic ions is maintained in solution for long periods, linked to the chelating agent, avoiding precipitation. In alternative, ethylenediaminetetraacetic acid (EDTA), whose molecular formulae is $(CH_2N)_2(CH_2COOH)_4$, is another chelating agent that equally forms stable metal complexes, which has determined its use as titrant for the determining of the concentration of several metal ions, in complexometric titrations [50]. The use of EDTA, as chelating agent, was profusely used in the preparation of superconductors and dielectric materials [21, 54], but few studies were presented using EDTA in the preparation of CCTO gels. One of the most important aspects to consider when producing a homogeneous resin, without precipitations, is the stability of the complex chelate-metal ion. EDTA, in the non ionized form can be represented as H_4Y. In the totally dissociated form (Y^{4-}) the EDTA molecule can form until six covalent links with each metal ion, although each EDTA molecule can only complex with one metal ion. The reaction between totally dissociated EDTA (Y^{4-}) and the metal ion (M^{n+}) can be expressed as follows [50]:

$$Y^{4-} + M^{n+} \leftrightarrow MY^{n-4} \tag{13}$$

In this conditions, the formation constant (K_f) of the reaction is expressed as:

$$K_f = \frac{[MY^{n-4}]}{[M^{n+}][Y^{4-}]} \tag{14}$$

Although complexation process using EDTA is similar to the one observed at Pechini method (that uses citric acid as chelating agent), metal-EDTA complexation reactions present, in general, higher K_f values than those evolving metal-citric acid, which means a major stability of the metal-EDTA complexes.

Table 2 presents the values of the formation constant to the metal complexes formed from the constitutive ions of CCTO: Ca^{2+}, Cu^{2+}, Ti^{4+}.

Meanwhile, once EDTA is a weak polyprotic acid and only effectively reacts when completely dissociated, the pH of the solution affects the equilibrium of the complexation reaction (13). This fact justifies the addition of ammonia (NH_4OH), increasing the pH of the solution, facilitating the dissolution of EDTA and the consequent increase of Y^{4-} concentration.

Besides the regulation of pH, the addition of NH_4OH produces, with the NO_3 groups from the other organometallic precursors, ammonia nitrate (NH_4NO_3), which promotes thermal decomposition and allows the decrease of the calcination temperature needed to obtain CCTO [51, 23].

Considering these factors, we would say that solutions with higher pH contribute for the stabilization of the metal-EDTA complex and lower calcination temperatures. However, the addition of ammonia solution makes the calcination of the resin so exothermic that tends to promote the increasing of the mean size of calcined powder particles, and in certain conditions, can cause the loss of material due to the projection of material as a consequence of violent emission of combustion gases. To avoid these effects it is convenient to optimize the use of ammonia, adjusting the pH of final solution to a minimum value.

To obtain the pH value that better serves to obtain a stable resin, we calculated the conditional formation constant (K_{cf}) for the several complexes to consider during the production of CCTO: Ca-EDTA, Cu-EDTA e Ti-EDTA. According to the K_{cf} values presented at the table 3, the complex Ti-EDTA is the one with lower stability along the pH scale, and a significant increase of K_{cf} is observed for $6 < pH < 7$. So, during the final phase of the preparation of the resin the pH of the solution was adjusted around those values, and the temperature increased to 80 °C, which would allow the production of stables metal-EDTA complexes.

Table 2. Formation constants (K_f) of the complexation reactions of EDTA-metal and citric acid-metal [46, 34]

	EDTA, log K_f	Citric Acid, log K_f
Ca^{2+}	10.69	3.5
Cu^{2+}	18.80	6.1
Ti^{4+}	3.43	5.4

Table 3. Conditional formation constants of the M-EDTA complexes, as a function of pH, at room temperature [46, 34]

pH	Ca-EDTA	Cu-EDTA	Ti-EDTA
10	1.71E+10	2.21E+18	9.42E+02
9	2.55E+09	3.28E+17	1.40E+02
8	2.64E+08	3.41E+16	1.45E+01
7	2.35E+07	3.03E+15	1.29E+00
6	1.08E+06	1.39E+14	5.92E-02
5	1.71E+04	2.21E+12	9.42E-04
4	1.76E+02	2.27E+10	9.69E-06

2. Preparation of the CCTO Powders

The CCTO powders were prepared by two different wet-chemical methods, which will be here designated by ethylenediaminetetracetic acid-gel method (EDTA-gel) [5] and modified Pechini (PM) process [6]. Attending to the similarities between these methods we will describe in the same section the several stages of each, pointing, whenever needed, the main differences among them. The preparation of the precursors was done using the following starting reactants: $Ca(NO_3)_2.4H_2O$ (≥ 99,0%, *Aldrich*, Germany), $Cu(NO_3)_2.3H_2O$ (98,0-103%, *Fluka*, Germany), $Ti(OC_4H_9)_4$ (butyl titanate, 97%, *Aldrich*, Germany), EDTA (≥ 99,0%, *Fluka*, Germany), 25% ammonia solution (*Fluka*, Germany) and 60% nitric acid (*J.M.G. Santos*, Portugal). For the PM method it was also used ethylene glycol (*Aldrich*, Germany), which will promote the esterification reaction between the carboxylic groups from EDTA molecules and the hydroxyl groups from ethylene glycol.

For both methods, in a first step, EDTA was dissolved in a minimum portion of ammonia solution to allow the complete dissolution of EDTA. After homogenization, it was gradually added an equimolar quantity of $Ti(OC_4H_9)_4$ relative to the amount of used EDTA. The solution was stirred at 80 °C after the addition of some drops of nitric acid, to adjust the pH to 4, facilitating the dissolution of $Ti(OC_4H_9)_4$ [54]. After one hour of stirring, the solution becomes yellowish and translucent, ready to be transferred to the stove, where it remained at 80 °C. In a second beaker, with EDTA dissolved as above, $Cu(NO_3)_2.3H_2O$ and $Ca(NO_3)_2.4H_2O$ were gradually added with continuous stirring so that we could obtain a solution with a molar ratio equal to Ca^{2+}: 3 Cu^{2+}: 4 EDTA. To avoid Cu precipitation, the pH of the solution was adjusted to 6.5 by further addition of ammonia. The final solution, which resulted from the mixture of the both previously described ones, was stirred continuously for 1 h at 80 °C. In the EDTA-gel method, the dark blue final solution (figure 9a) was transferred to an oven at 80 °C, where it remained for 2 days, for the dehydrating step. After this period, no signs of precipitation were detected which can be a proof of the chemical stability of the formed complexes. In the PM method, an equimolar quantity of ethylene glycol, relative to the amount of EDTA, was added to the mixture. The resulting mixture was vigorously homogenized during 1 h at 80 °C. After, it was placed in the oven at 80 °C for 2 days. This blue resin did not show also any signs of precipitation (figure 9 b). Based on the interpretation of the differential thermal analysis (DTA), both resins were calcined directly in a furnace at 600 °C, 700 °C, 800 °C and 900 °C, during 2 h. Each of the calcined powders will be identified with the name of the method and the calcination temperature. As an example, EDTA_C800 and PM_C800 refer to the powders calcined at 800 °C, obtained from EDTA-gel and PM resins, respectively. The obtained brown powders were grounded using mortar and pestle and uniaxially pressed into discs with 7 mm in diameter and 1 mm in thickness, by applying a pressure of 2500 bar. Sintering process was performed in a conventional furnace at 1050 °C for 12 h. The flow chart, presented at figure 10, resumes the main steps for the preparation of the resins.

(a) (b)

Figure 9. Resins obtained by a) EDTA-gel method and b) modified Pechini method.

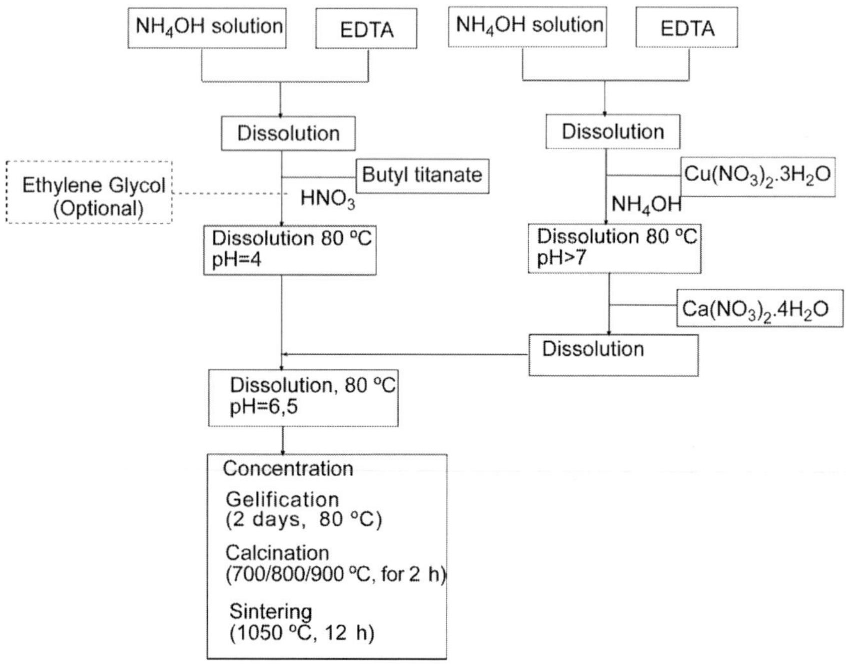

Figure 10. Flow chart for the sol-gel synthesis of CCTO.

3. Characterization of the Samples

3.1. Thermal Decomposition of the Resin and Formation of Products

The results of differential thermal analysis (DTA), X-ray diffraction (DRX) and scanning electron microscopy (SEM) were used to study the structural and phase changes that occur in the resin during the calcination process.

Figure 11 presents the result of DTA analysis, in the temperature region between 20 °C and 1100 °C, applying a heating rate of 5 K/min, for EDTA-gel and modified Pechini (PM) resins, respectively, previously submitted to heat treatment at 300 °C. Analyzing the DTA results of EDTA-gel resin it is possible to point four main stages during the decomposition process. Between 25 °C and 300 °C, is identified an endothermic band, with a peak at 85 °C, which can be related with the loss of free water linked to the gel. Second stage, appears beyond 200 °C, where two exothermic phenomena are identified. The first exothermic phenomenon, between 200 °C and 600 °C, with a peak at 343 °C, should be related with the breaking of metal-EDTA bonds and consequent liberation of H_2O, CO_2 and nitrogenous compounds (N_2, NO_2). After this step, all the NH_4NO_3 should be decomposed, which is confirmed by the XRD analysis to the 600 °C calcined powders (figures 12 and 13). The peak at 343 °C observed in the DTA corresponds to the burnout of organic compounds of the precursor powders. The phenomenon centered at 683 °C can be attributed to the beginning of crystallization process, which is corroborated by the XRD analysis, once only powders calcined at more than 700 °C show the presence of CCTO crystal phase (figure 12).

From DTA analysis obtained for the PM resin (figure 11) it is possible to identify five phenomena during the decomposition process. The two first phenomena are consistent with those observed in the decomposition of the EDTA-gel resin. A new exothermic phenomenon, with a peak centered at 410 °C, inexistent at the EDTA-gel resin DTA, is observed. This transformation can be related to the breaking of strong polymeric chain bonds. The exothermic peak at 750 °C can be attributed to the beginning of the CCTO crystallization.

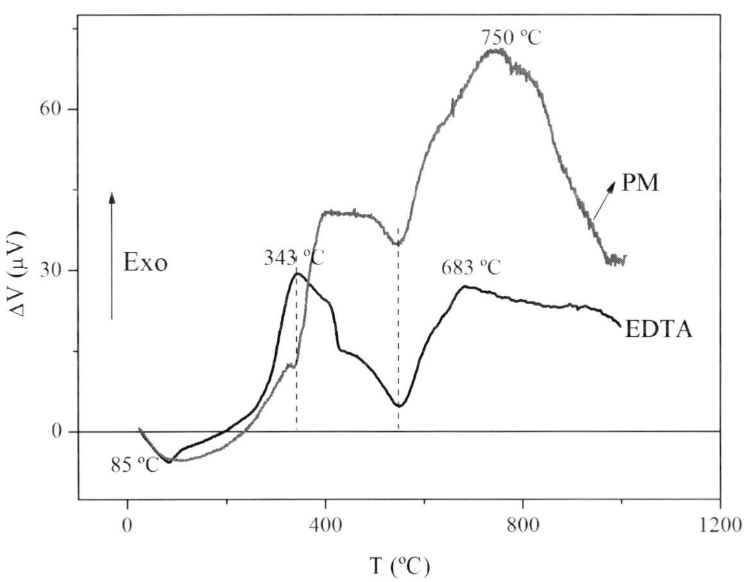

Figure 11. Differential thermal analysis of the resins obtained by EDTA-gel and PM method.

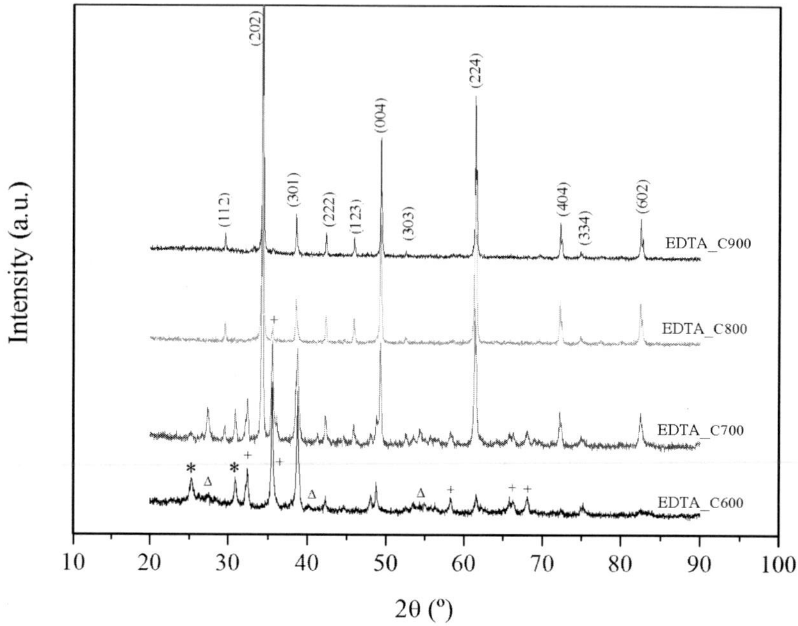

Figure 12. XRD spectrum of the powders obtained from the calcined resins at 600 °C (EDTA_C600), 700 °C (EDTA_C700), 800 °C (EDTA_C800) and 900 °C (EDTA_C900). Secondary phases: + CuO, Δ TiO_2, * non identified.

Figure 12 shows the XRD patterns of the powders obtained from the calcination of the EDTA-gel resin at four different temperatures: 600 °C (EDTA_C600), 700 °C (EDTA_C700), 800 °C (EDTA_C800) and 900 °C (EDTA_C900). It is observed that CCTO crystal phase (JCPDS N° 75-2188) is present in all the samples, except on the EDTA_C600. However, particularly at lower calcination temperatures (< 800 °C), it can be found secondary phases, as CuO and TiO_2. For temperatures higher than 800 °C only residual portions of CuO are detected.

Figure 13 presents the XRD patterns of the PM powders calcined at 700 °C (PM_C700), 800 °C (PM_C800) e 900 °C (PM_C900). XRD patterns obtained for all the samples point CCTO as the major phase, while CuO is also present but as a minor phase. For lower calcination temperatures (PM_C700), some TiO_2 residues were also detected.

Table 4 summarizes the results of crystallite size (L_c), for both series, calculated using Scherrer formula [43]. The results show that L_c increases with the increasing of calcination temperature. The cell parameters, obtained from XRD patterns, were refined using CELREF program [28]. The values calculated for the *a* parameter are in accordance with the standard JCPDS: 75-2188 (7.391±0.001 Å).

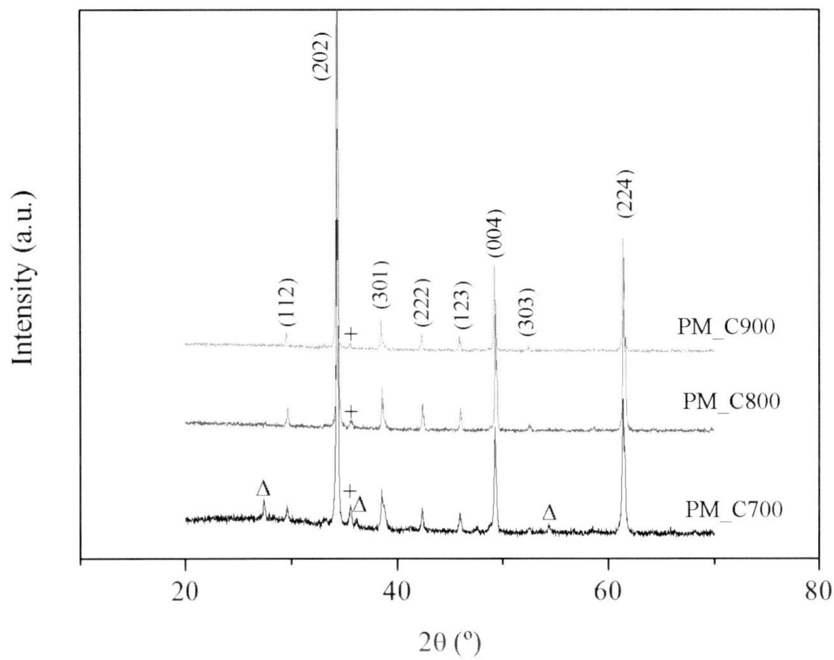

Figure 13. XRD spectrum of the powders obtained from the PM calcined resins at 700 °C (PM_C700), 800 °C (PM_C800) and 900 °C (PM_C900). Secondary phases: + CuO, Δ TiO_2.

Table 4. Crystallite size (L_c) and cell parameter (a) of the calcined powders

Sample	L_c (nm)	Cell parameter a (Å)
EDTA_C700	67 ± 6	7.400 ± 0.004
EDTA_C800	87 ± 8	7.398 ± 0.002
EDTA_C900	145 ± 15	7.396 ± 0.003
PM_C700	72 ± 6	7.390 ± 0.006
PM_C800	131 ± 14	7.417 ± 0.009
PM_C900	211 ± 19	7.409 ± 0.005

Figure 14. SEM micrographs of the calcined powders: (a) EDTA_C700, (b) EDTA_C800 and (c) EDTA_C900.

3.2. Morphologies of Calcined Powders

Figure 14 shows the micrographs of the powders prepared by EDTA-gel technique. All micrographs show the formation of nanoparticles that gradually change their shape with the increasing of the calcination temperature. Particles of the calcined powders at 700 °C (EDTA_C700) present spherical shape, with dimensions between 50 and 100 nm. The micrographs of EDTA_C800 and EDTA_C900 samples show the formation of necks between neighbor particles, which, besides the growing of their dimensions, relatively to EDTA_C700, contributes to the formation of elongated shapes.

Figure 15 shows the SEM micrographs of the powders prepared by PM method, after the calcination process.

The particles of the powder calcined at 700 °C (PM_C700) present a spherical shape, with dimensions between 100 and 200 nm. In the micrographs of the PM_C800 and PM_C900 samples it was detected, as in the samples prepared by the EDTA-gel method, the formation of necks between the initial touching particles, which cause the growth of CCTO elongated particles. As a consequence the particles of the samples calcined at 900 °C show an average size between 500 nm and 1 μm.

3.3. Structural Analysis of the Sintered CCTO

The sintered pellets were characterized by XRD and energy dispersive spectroscopy (EDS). Figure 16 shows the XRD patterns of the samples EDTA_S700, EDTA_S800 and EDTA_S900, that result from the sintering of the calcined powders at 700 °C, 800 °C and 900 °C, respectively. For all the studied samples CCTO appears as the major phase. Only small quantities of CuO were detected.

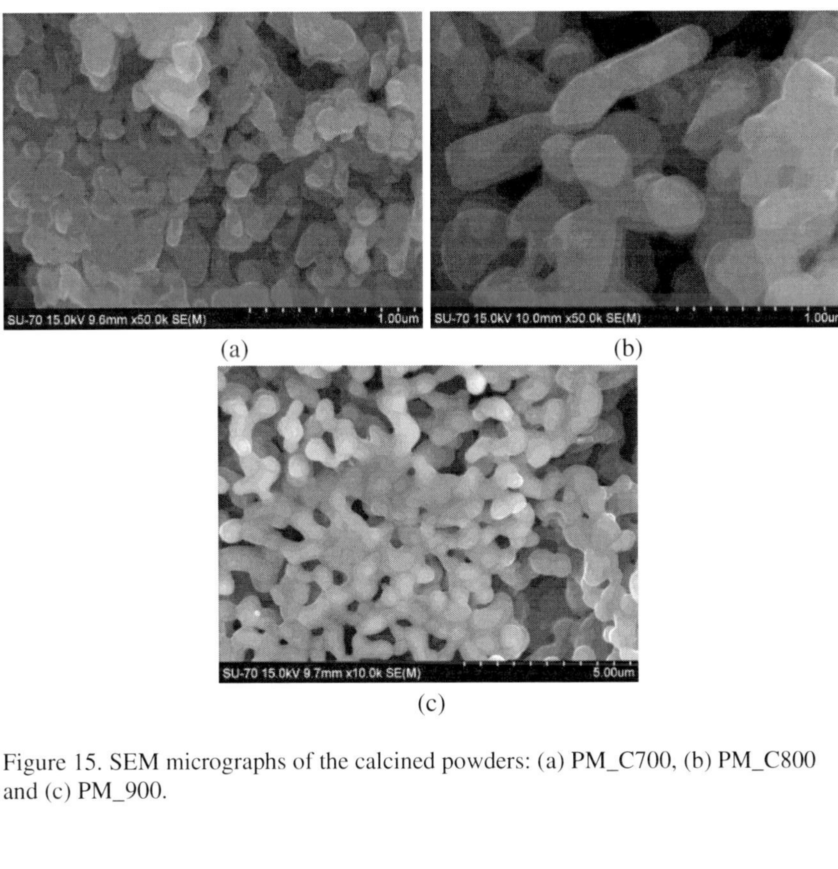

Figure 15. SEM micrographs of the calcined powders: (a) PM_C700, (b) PM_C800 and (c) PM_900.

EDS analysis (figure 17) shows that sintered pellets present a stoichiometric ratio for the cations Ca/Cu/Ti that slightly diverges from pure CCTO stoichiometry (1/3/4). For all the samples it is detected a deficiency of Cu, in the grains, comparatively to the other elements (Ca and Ti). It can be seen that for the sample EDTA_S800 the cationic ratio Ca/Cu/Ti, at the center of the grain (point f), is 1.0/2.7/3.9. This divergence, relatively to pure CCTO, is even more pronounced at the grain borders of the grain (point g), where the cationic ratio is 1.0/2.5/3.6. Near the boundaries of the grain it is also possible to observe a segregation area (arrow e) whose composition is rich in Cu.

Figure 18 shows the XRD patterns of PM_S700, PM_S800 and PM_S900, which result from the sinterization of the powders calcined at 700 °C, 800 °C and 900 °C, respectively. The CCTO crystal phase is identified as the major phase for all the samples, although some residues of CuO were also detected.

The EDS analysis performed to the pellets of this samples show that the proportion between the cations Ca/Cu/Ti present several differences relatively to the stoichiometry of pure CCTO. Measurements at the central area of the grains show an excess of Cu relatively to the other elements (Ca and Ti), which is different from the samples produced by the EDTA-gel method.

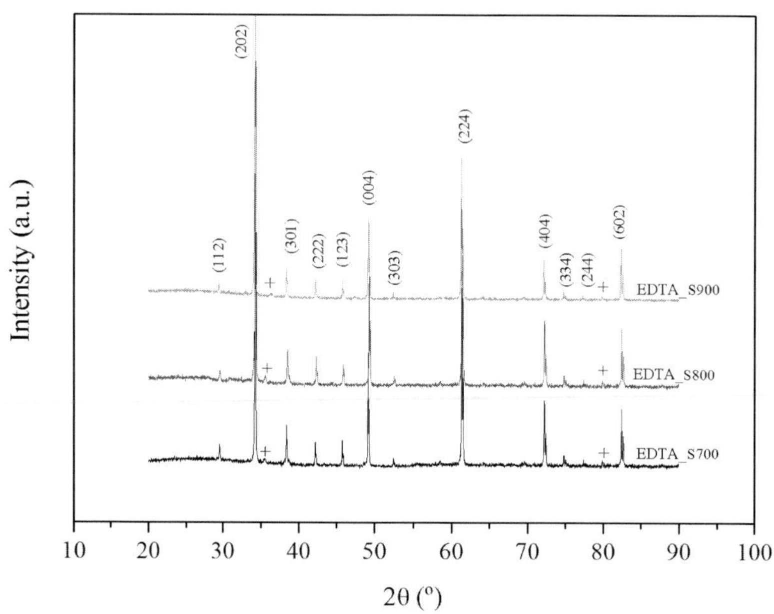

Figure 16. XRD spectrum of the sintered samples: EDTA_S700, EDTA_S800 and EDTA_S900. Secondary phases: + CuO.

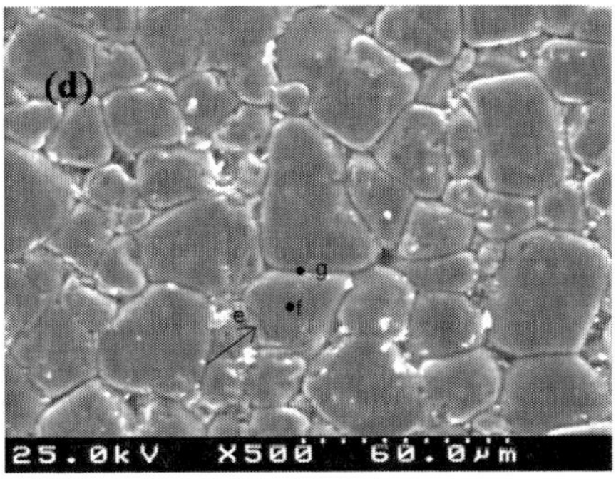

Point e – Segregation zone

Element	[n. at.-%]	[n. X/n. Ca]
Ca K	3.89 ± 0.88	1.0
Cu K	80.14 ± 11.60	20.6
Ti K	15.97 ± 2.07	4.1

Point f – Grain

Element	[n. at.-%]	[n. X/n. Ca]
Ca K	13.15 ± 1.54	1.0
Cu K	35.70 ± 4.90	2.7
Ti K	51.15 ± 5.04	3.9

Point g – Grain boundary

Element	[n. at.-%]	[n. X/n. Ca]
Ca K	14.16 ± 1.74	1.0
Cu K	34.93 ± 5.02	2.5
Ti K	50.90 ± 5.22	3.6

Figure 17. EDS analysis at several points of the free surface of the EDTA_S800 sample.

It can be seen in figure 19 that, for the sample PM_S700, the cationic ratio Ca/Cu/Ti at the centre of the grain is 1.0/3.6/3.8. However, an analysis performed in points close to the boundaries of the grain show a decrease of the Cu content. In those points the Ca/Cu/Ti cationic ratio is 1.0/2.9/4.4, which can be justified by the loss of Cu next to the boundaries of the grains, used in the formation of segregations around the grains, as it is observed in the figure 19. The EDS analysis in the segregation area of the PM_S700 samples proves that the identified phases is rich in Cu and have a deficit of Ca and Ti.

3.4. Microstructure of the Sintered CCTO

Figures 20 and 21 show the microphotographs of the free surfaces obtained from the sintered pellets of EDTA-gel and PM samples, respectively. Although all the samples have been submitted to the same sintering conditions, the pellets surface present microstructural differences, particularly their average grain sizes, which imply that the calcination process of the CCTO powders strongly influences the microstructure of the sintered samples.

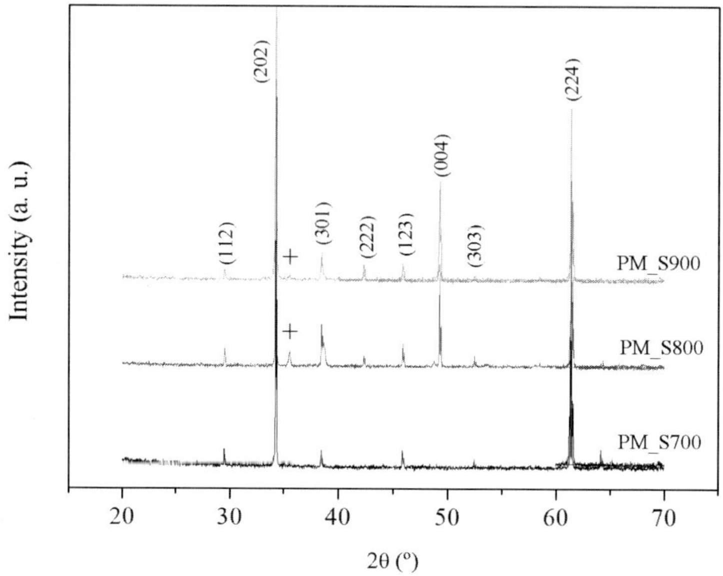

Figure 18. XRD spectrum of the sintered samples: PM_S700, PM_S800 and PM_S900. Secondary phase: + CuO.

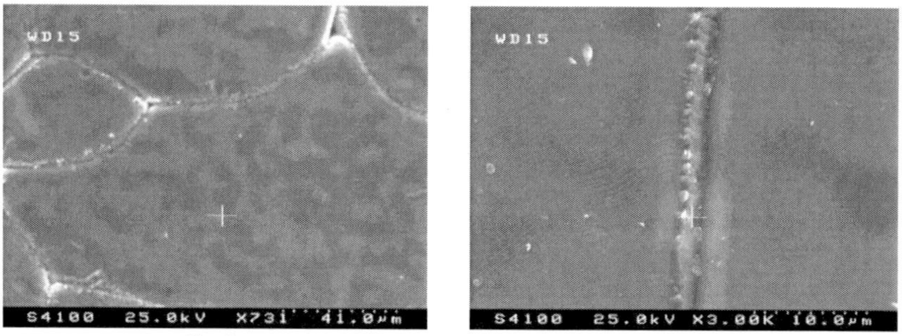

Grain

Element	[n. at.-%]	[n. X/n. Ca]
Ca K	**12.02 ± 1.83**	1.0
Cu K	**42.75 ± 9.18**	3.6
Ti K	**45.23 ± 6.24**	3.8

Grain boundaries

Element	[n. at.-%]	[n. X/n. Ca]
Ca K	**5.41 ± 0.85**	1.0
Cu K	**73.45 ± 14.19**	13.6
Ti K	**21.13 ± 2.79**	3.9

Figure 19. EDS analysis at several points of the free surface of the PM_S700 sample.

Using the computational program *Image J*, more than 30 grains, for each sample, were inspected. After statistical treatment of these data, average grain size distribution results were obtained (table 5).

It is found that the average grains size of the PM_S700 presents the major value among the studied samples, being significantly higher than the value measured for the samples prepared by the EDTA-gel method. The other samples, PM_S800 and PM_S900, present a significant decrease of the average grain sizes comparatively to the sample PM_S700, which makes difficult to establish a correlation between the size of the calcined powders and the grain size of the sintered samples.

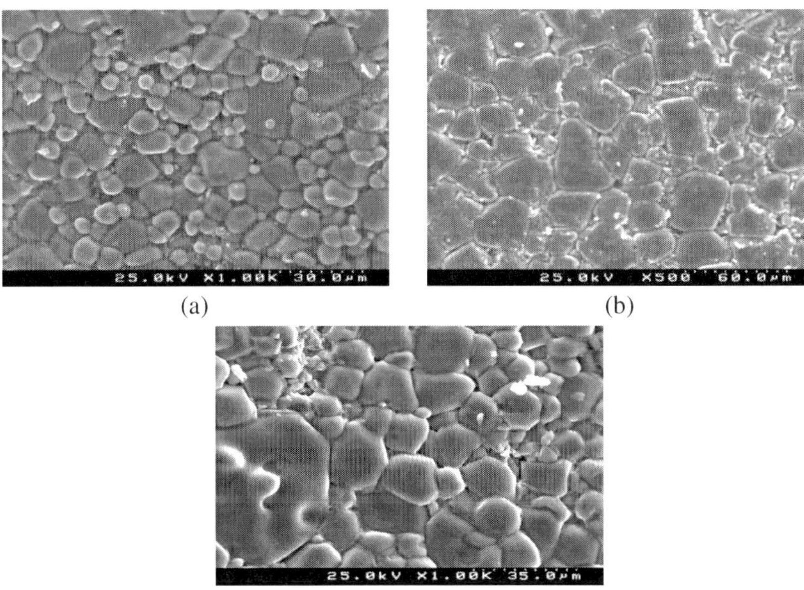

(a) (b) (c)

Figure 20. SEM micrographs of the free surfaces of the sintered samples: (a) EDTA_S700, (b) EDTA_S800 and (c) EDTA_S900.

Table 5. Average grains size of the sintered pellets

Calcination temperature (°C)	Average grains size (µm)	
	EDTA samples	PM samples
700	7 ± 4	84 ± 51
800	15 ± 8	36 ± 22
900	14 ± 8	8 ± 6

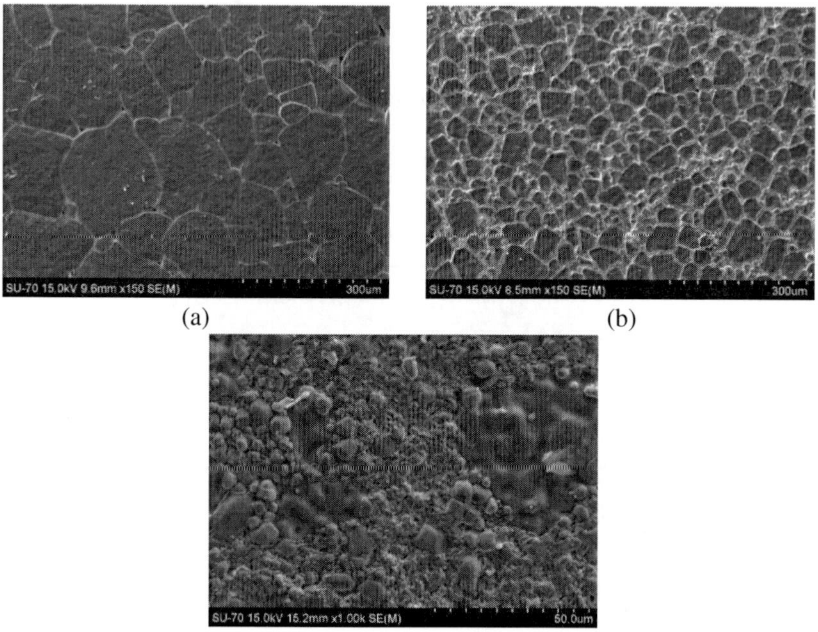

(a) (b)

(c)

Figure 21. SEM micrographs of the free surfaces of the sintered samples: (a) PM_C700, (b) PM_C800 e (c) PM_900.

3.5. Dielectric Measurements

Electrodes were made by painting with conductive silver paint the both faces of all studied pellets. Figures 22 and 23 represent the real (ε') and imaginary (ε'') parts of the complex permittivity ($\varepsilon^* = \varepsilon' - j\varepsilon''$), at room temperature, in function of frequency (*f*), in the frequency range 40 Hz – 30 MHz, and applying a voltage of 500 mV, for EDTA-gel and PM series, respectively.

For both series, the permittivity vs. frequency shows the existence of a relaxation process in the MHz region, generally associated to a Maxwell-Wagner effect, that can be ascribed to the different conductivities between grains and grain boundaries. Besides this relaxation, it is possible to identify a second relaxation, in the kHz region, originated by the formation of depletion layers at the electrode/samples surface/bulk interfaces [krohn08]. The existence of these two relaxations is confirmed by the observation of two semicircles at the Cole-Cole representation of the complex permittivity (figures 24 and 25). In these figures it is also visible a component that can be associated to the d.c. conductivity, with the shape of a line segment in the region of the low frequencies.

The high dielectric constant measured for these samples ($\varepsilon' \approx 10^4$), in the low frequency region, is comparable with the results obtained by other authors, like Masingboon et al. [35], Jin et al. [25] e Liu et al. [31], who also prepared CCTO using other wet-chemical methods. At room temperature, it is possible to observe, for all the samples, that ε' abruptly drops to values around 100, for all the frequencies higher than 10 MHz (figures 22 and 23).

Table 6 presents the results of the fitting of Cole-Cole function to the high frequency region complex permittivity data, at room temperature. Considering EDTA-gel series, maximum dielectric strength value ($\Delta\varepsilon = 7873$) is achieved for the EDTA_S900 sample.

Regarding PM series, we verify that the maximum dielectric strength is measured for the PM_S700 sample ($\Delta\varepsilon = 13722$). It is also noticed that, although the other PM samples present lower $\Delta\varepsilon$ values than the PM_S700 sample, their values are higher than the $\Delta\varepsilon$ measured for the samples prepared by the EDTA-gel method. The parameter α results obtained for all the PM and EDTA samples show that the relaxation is almost Debye type ($\alpha \sim 0$).

Considering that the behavior of the CCTO ceramics can be modeled as the series combination of two parallel RC circuits, one representing the grain effects (low resistive component) and the other the grain boundaries (high resistive component), the dielectric relaxation time, τ, is given by:

$$\tau = \frac{R_g R_{gb}\ C_g + C_{gb}}{R_g + R_{gb}} \tag{15}$$

where C_g and R_g are, respectively, the capacitance and resistance of the grains and C_{gb} and R_{gb} are the capacitance and resistance of the grain boundaries. Considering that $R_{gb} >> R_g$ and $C_{gb} > C_g$ the expression (15) can be simplified, resulting the following:

$$\tau \approx R_g C_{gb} \tag{16}$$

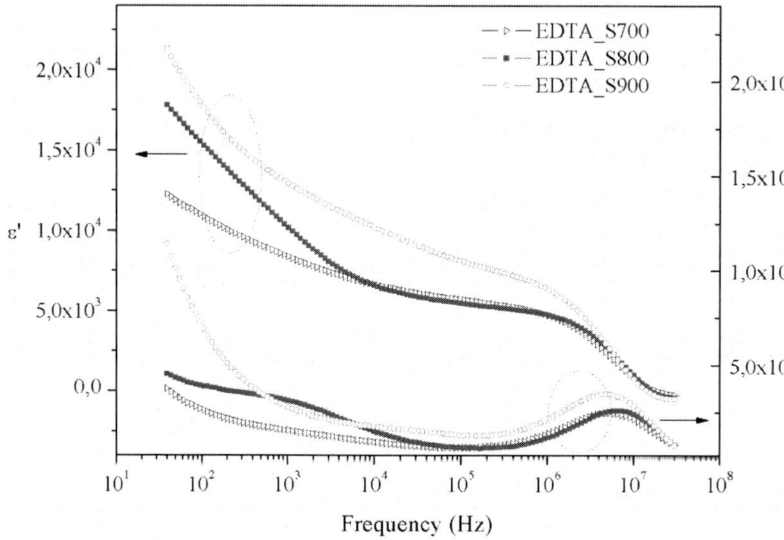

Figure 22. Real (ε') and imaginary (ε'') parts of the complex permittivity (ε*), for the EDTA-gel sintered samples, at room temperature.

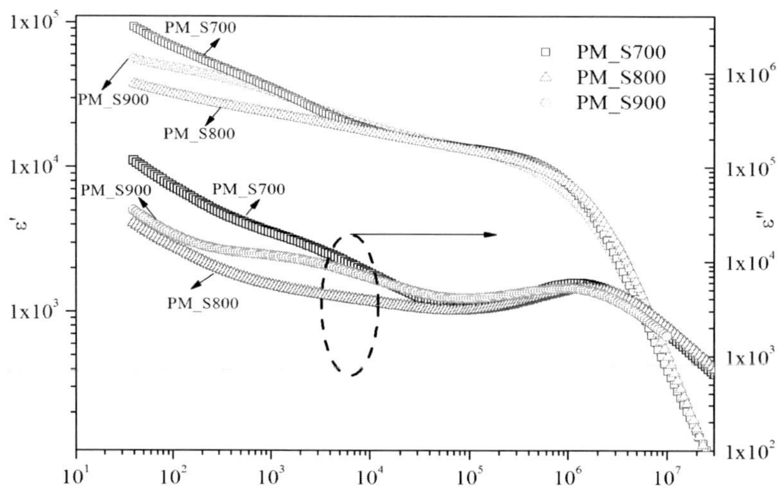

Figure 23. Real (ε') and imaginary (ε'') parts of the complex permittivity (ε*), for the PM sintered samples, room temperature.

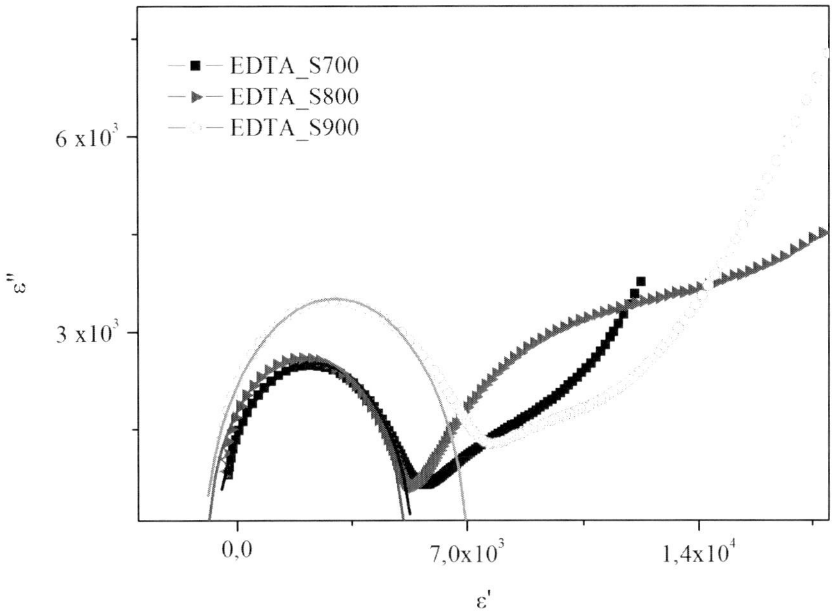

Figure 24. Cole-Cole plot of the complex permittivity, for the EDTA-gel sintered samples, at room temperature.

The values of τ obtained for the PM samples (table 6), resulting from the fitting with the Cole-Cole function to the ε^* data, are higher than those measured for the samples prepared by the EDTA-gel method. These results agree with the predictions of expression (16), once that R_g values measured for the PM samples are higher than those measured for the EDTA-gel samples.

Table 6. Relaxation parameters, grains resistance (R_g) and grain boundaries capacitance (C_{gb}), obtained for the sintered samples, at room temperature

Sample	τ (s)	f_{max} (MHz)	$\Delta\varepsilon$	α	R_g (Ω)	C_{gb} (nF)
EDTA_S700	$3.13x10^{-8}$	5.08	6452	0.17	10	3.04
EDTA_S800	$2.43x10^{-8}$	6.55	6488	0.14	13	3.56
EDTA_S900	$3.32x10^{-8}$	4.79	7873	0.14	9	4.47
PM_S700	$1.42x10^{-7}$	1.12	13722	0.09	100	1.17
PM_S800	$1.06\ x10^{-7}$	1.50	12124	0.09	90	1.07
PM_S900	$1.25\ x10^{-7}$	1.27	12858	0.06	76	1.51

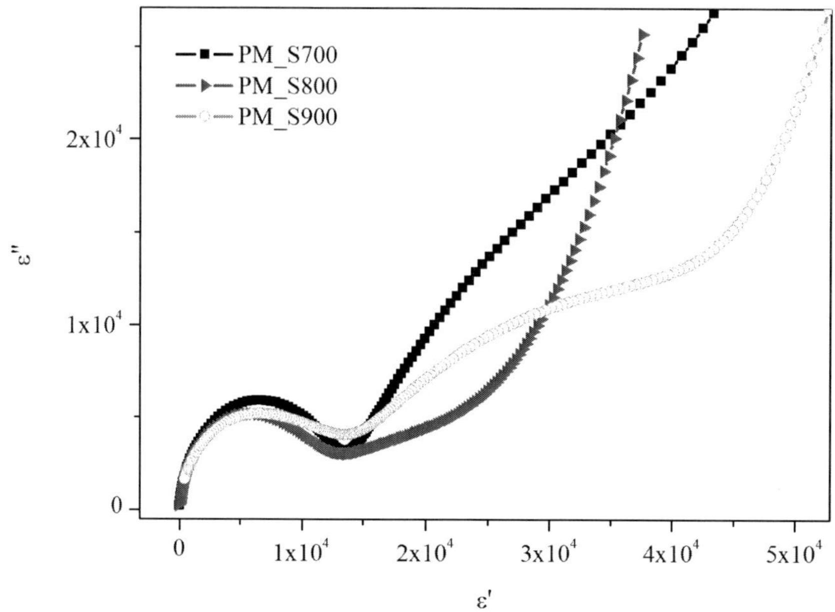

Figure 25. Cole-Cole plot of the complex permittivity, for the PM sintered samples, at room temperature.

Adjusting the measuring temperature to 100 K, for the PM samples, it is observed a decrease of ε_s, reaching ε_∞ a value around 100, for all the samples (figure 26). This implies that, as illustrated at table 7, $\Delta\varepsilon$ decreases.

However, decreasing the measuring temperature, the low frequency relaxation is no longer observed, so that only a semicircle in the Cole-Cole representation of ε^* is presented (figure 27). Also the component associated to the d.c. conductivity is minimized.

The results of the fitting, using the Cole-Cole function, are presented in table 7. It should be emphasized that, although the lower measuring temperature, $\Delta\varepsilon$ still presents high values (> 5000) for all the samples. The relaxation time (τ) shows a significant increase for all the samples, which is a consequence of the strong increase of R_g ($\tau = R_g\ C_{gb}$). It can be also observed an increase of α with the decrease of temperature, which can be attributed to an increase of the dipolar interaction.

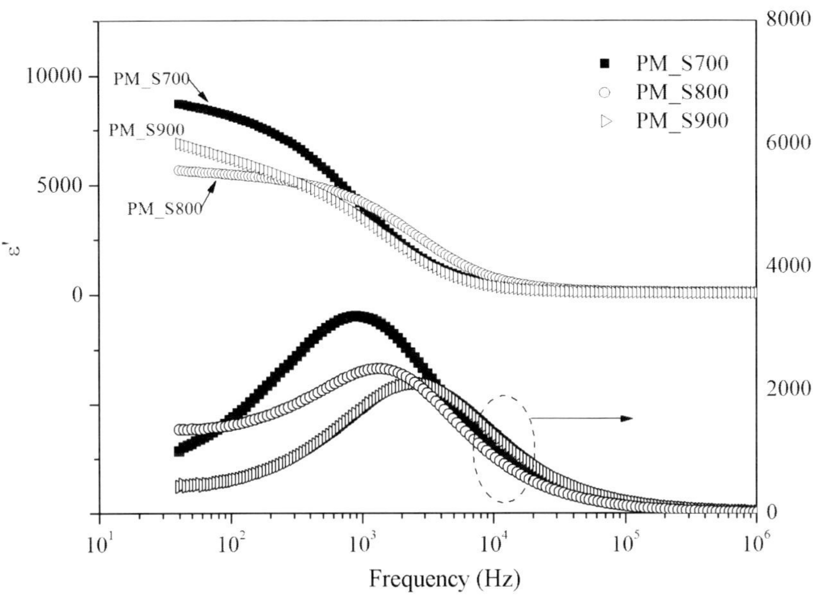

Figure 26. Real (ε') and imaginary (ε'') parts of the complex permittivity (ε*), obtained for the PM sintered samples, measured at 100 K.

Table 7. Relaxation parameters, grains resistance (R_g) and grain boundaries capacitance (C_{gb}), for the sintered samples, at 100 K

Samples	$\Delta\varepsilon$	τ (10^{-4} s)	α	R_g (kΩ)	C_{fg} (nF)
PM_S700	8730	1.77	0.19	152.0	0.743
PM_S800	5460	0.58	0.15	838.9	0.484
PM_S900	6130	1.22	0.16	123.5	0.718

Figures 28 and 29 present the dielectric loss (tan δ), as a function of frequency, at room temperature, for EDTA-gel and PM series, respectively. Dielectric losses are high (> 0.1) in all frequency range and for all the samples. This high dielectric losses can be associated to an intense d.c. conductivity component that influences the low frequency region. This dielectric loss component can be considerably decreased when decreasing the measuring temperature.

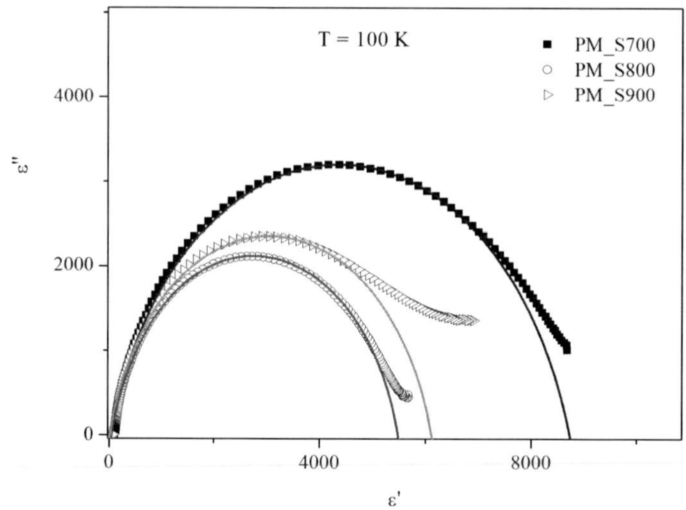

Figure 27. Cole-Cole plot of the complex permittivity, for the PM sintered samples, at 100 K.

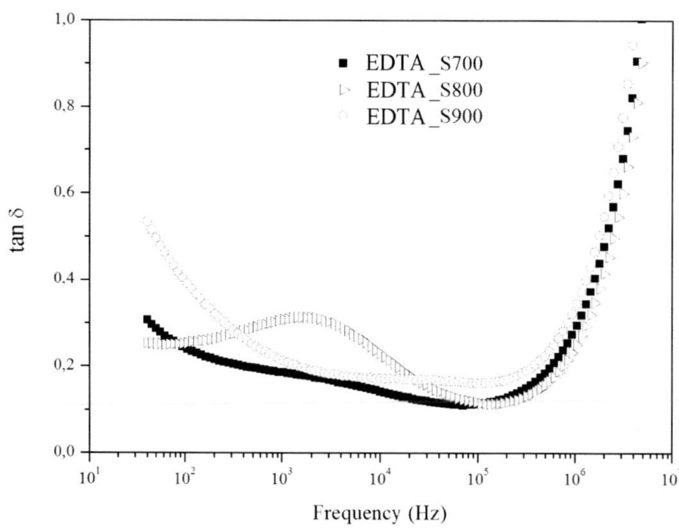

Figure 28. Dielectric loss (tan δ), for the EDTA-gel sintered samples, at room temperature.

Decreasing the measuring temperature until 100 K, the influence of the d.c. conductivity response on the tan δ is strongly reduced (figure 30), so that at 40 Hz the values of tan δ measured for PM_S700, PM_S800 and PM_S900 are equal to 0.117, 0.079 and 0.198, respectively.

At this temperature (100 K), the Maxwell-Wagner relaxation presents a relaxation frequency in the kHz region, which determines the increase of tan δ at this spectral region.

After this relaxation, tan δ decreases to values close to those at the pre-relaxation region, so that at 10 MHz the tan δ values measured for PM_S700, PM_S800 and PM_S900 are equal to 0.068, 0.091 and 0.075, respectively.

3.5.1. High Frequency Relaxation

Figure 31 shows the results of complex admittance, at room temperature, obtained for the sintered EDTA-gel pellets.

The fitting of these data was performed using Cole-Cole function [11]:

$$Y^* = \frac{Y_g}{1 + \left(i\omega\tau_g\right)^{1-\alpha}} \tag{17}$$

where Y_g represents the admittance of the grains, τ_g the relaxation time and α is a parameter between 0 and 1 associated to the dipolar interaction.

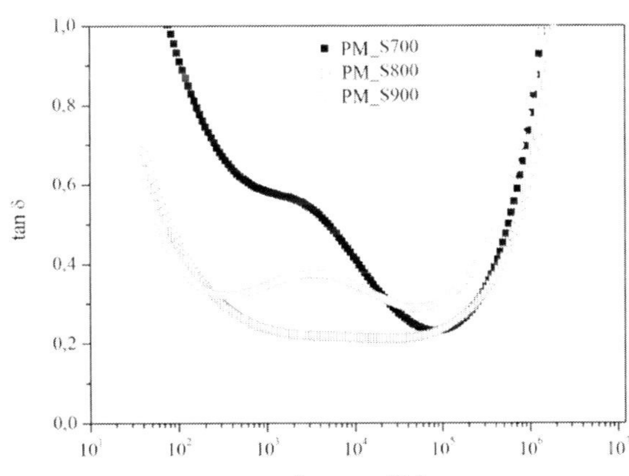

Figure 29. Dielectric loss (tan δ), for the PM sintered samples, at room temperature.

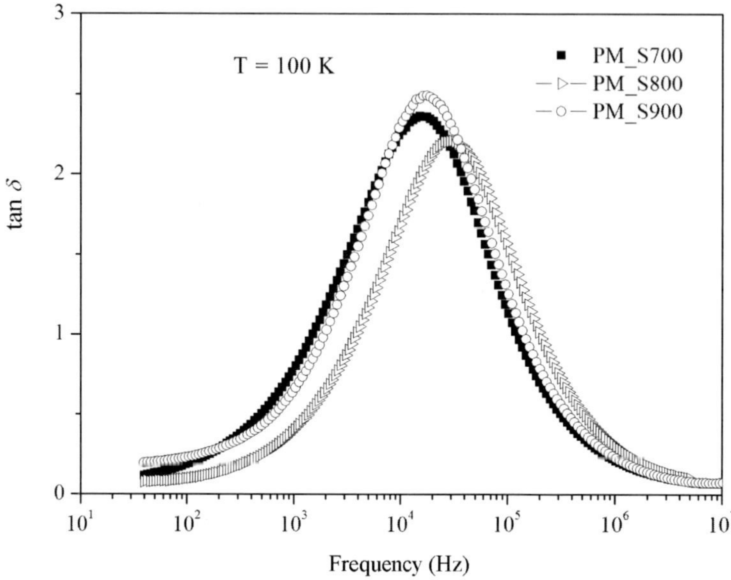

Figure 30. Dielectric loss (tan δ), for the PM sintered samples, at 100 K.

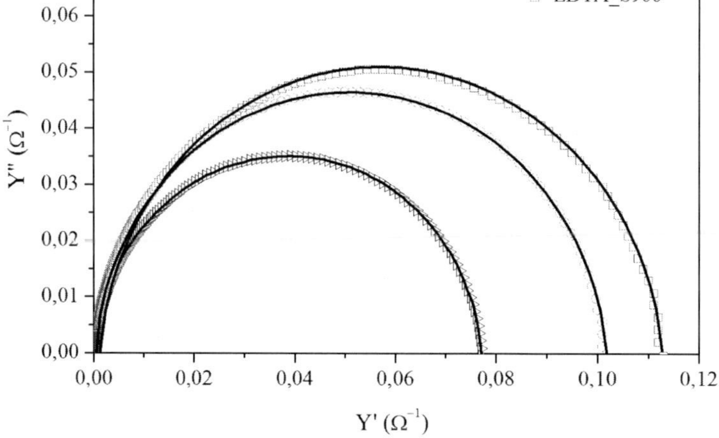

Figure 31. Nyquist representation of the complex admittance (Y^*), for the EDTA-gel sintered samples, at room temperature.

From the fitting of the experimental data with equation (17), the relaxation parameters were obtained, and are presented in table 8.

The analysis of this table points to a similar value of τ_g ($\tau_g \approx 10^{-8}$ s) for all the samples, which mean that at high frequencies the relaxation mechanism is the same, independently of the synthesis process. R_g values are very low for all the samples (< 20 Ω). Even though, the highest value was measured for the PM_S900 sample, corresponding to a corrected resistance ($R_g.A/L$) of 1.01 Ω.m. These low resistance values are usually associated to the grains response [45, 24] and reveal the semiconductor character of these structures, so that the increase of calcination temperature does not produce significant changes of the corrected resistance associated to the grains ($R_g.A/L$).

Analyzing the α parameter behavior, presented in table 8 it is found that for PM samples it increases with the increasing of calcination temperature. Contrarily to the behavior observed in the EDTA-gel samples, where α remains insensible to the different microstructures, the increase of the heterogeneity of PM samples microstructure, specially motivated by the increment of Cu segregation around the grains, can be the main reason for the increase of α.

3.5.2. Low Frequency Relaxation

At room temperature it is not possible to calculate, without a high uncertainty level, the equivalent resistance of the samples at low frequencies. Anyway, as it is exemplified at figure 32 for the EDTA-gel series, the resistance associated to this structure, generally accepted as associated to the grain boundaries response, is much higher (> 100 kΩ) than the grains resistance, which may be the reason for the arising of the Maxwell-Wagner effect.

Table 8. Relaxation parameters for the sintered samples, at room temperature

Sample	τ_g (10^{-8} s)	Y_g (Ω^{-1})	$R_g.A/L$ (Ω.m)	α
EDTA_S700	2.97	0.10	0.62	0.06
EDTA_S800	2.68	0.08	0.69	0.05
EDTA_S900	3.40	0.11	0.58	0.07
PM_S700	9.89	0.010	0.96	0.06
PM_S800	7.37	0.012	0.83	0.08
PM_S900	8.51	0.013	1.01	0.14

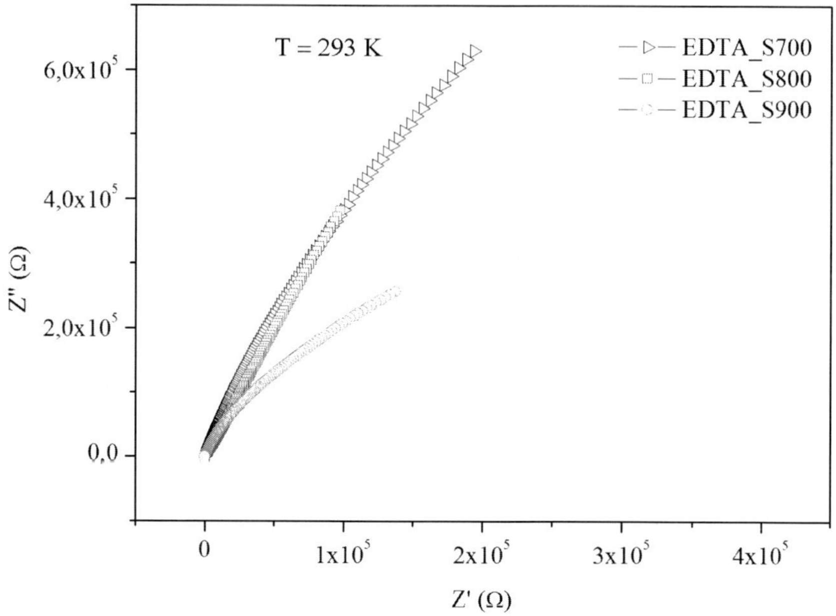

Figure 32. Nyquist representation of the complex impedance (Z^*), for the EDTA-gel sintered samples, at room temperature.

Increasing the temperature of measurement, the relaxation at the low frequency region becomes visible in the Nyquist representation of Z^* (Z'' vs. Z'), allowing the correct fitting of all the experimental data with the Cole-Cole function [11]:

$$Z^* = R_g + \frac{R_{gb}}{1 + \left(i\omega\tau_{gb}\right)^{1-\alpha_{gb}}} \tag{18}$$

Using equation (18), several fittings were performed for the series EDTA-gel and PM, at 393 K and 360 K, shown in figures 33 and 34, respectively. The fitting parameters are in table 9. The τ_{gb} parameter is, for all the samples, much higher than the respective value of τ_g (table 8), which can be justified with the existence of another relaxation mechanism, commonly accepted as associated to the grain boundaries.

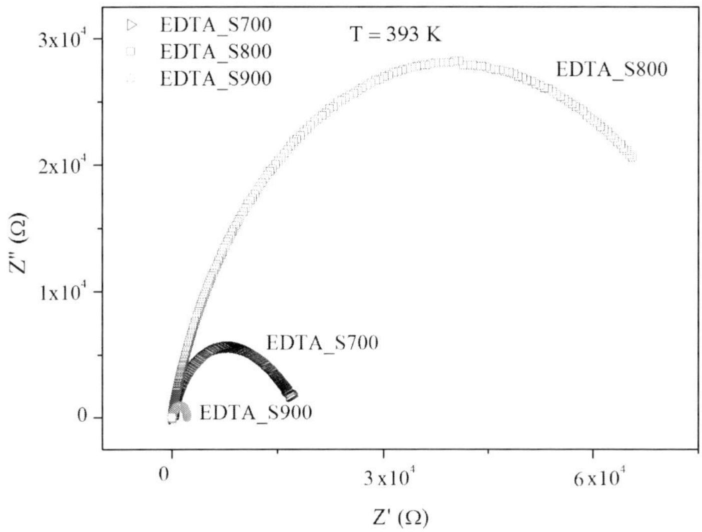

Figure 33. Nyquist representation of the complex impedance (Z^*), for the EDTA-gel sintered samples, at 393 K.

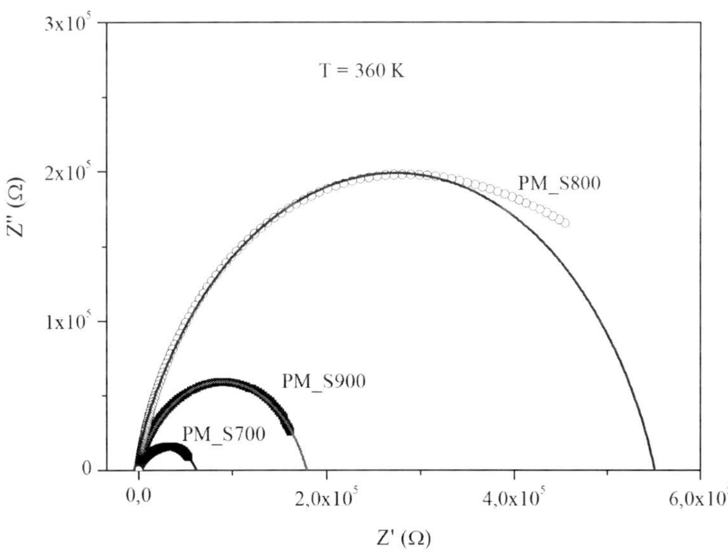

Figure 34. Nyquist representation of the complex impedance (Z^*), for the PM sintered samples, at 360 K.

Table 9. Relaxation parameters obtained for the second relaxation, at 360 K and 393 K. d_g represents the average grain size

T(K)	Sample	τ_{gb} (10^{-4} s)	$R_g .A/L$ (Ω.m)	$R_{gb}.A/L$ (kΩ.m)	α_{gb}	ε_s (10^4)	d_g
393	EDTA_S700	1.12	2.71	0.931	0.07	1.35	7
	EDTA_S800	9.76	1.98	4.87	0.08	2.26	15
	EDTA_S900	0.208	2.14	0.135	0.09	1.74	14
360[1]	EDTA_S700	20.4	0.94	18.7	0.20	1.23	7
	EDTA_S900	9.42	1.36	7.04	0.19	1.46	14
	PM_S700	4.30	0.506	0.591	0.38	8.22	84
	PM_S800	11.0	0.501	5.51	0.21	2.25	36
	PM_S900	4.64	0.105	2.33	0.26	2.25	8 ± 6

[1] It was not possible to obtain a proper fitting for the EDTA_S800 sample, at 360 K, because at this temperature the semicircle resulting from the Nyquist representation of Z* is very incomplete.

Even at higher temperatures it is observed a high conductivity difference between grains and grain boundaries. This fact should be related to the reoxidation of the grains surface during the sintering process, forming Schottky barriers, resulting from it a dielectric behavior similar to that of two back-to-back diodes [moulson, 1].

According to the Brick Layer model [mcdonald05], this type of microstructure causes the accumulation of charge within the grain boundaries, and consequently, a high number of these microstructures can explain the colossal dielectric constant measured for CCTO. Brick Layer model also predicts that static permittivity (ε_s) measured for the CCTO samples can be inferred with the following expression [30]:

$$\varepsilon_s = \varepsilon_{gb}\frac{(d_{gb} + d_g)}{d_{gb}}, \tag{19}$$

where ε_{gb} represents the dielectric permittivity of the grain boundaries, d_g is the mean grain size and d_{gb} is the thickness of the grain boundaries.

Once the measurement of d_{gb} is not easy, it was not possible to calculate ε_{gb}. However, if we suppose for ε_{gb} a value similar to the one of ε_g ($\approx$ 100), than the high values of ε_s ($>10^3$), measured for all the samples, are easily justified, attending that $d_g >> d_{gb}$.

Considering that $\omega_{max} = 1/\tau$, where $\omega_{max} = 2\pi f_{max}$ and $\tau = RC$, the values of ε_s ($\varepsilon_s = C_{gb}/C_0$) were estimated. As it is shown in table 9, higher values of ε_s

are found for the PM samples than for the EDTA-gel samples, confirming what was already pointed at room temperature. This fact is in agreement with the larger values of d_g measured for the PM samples relatively to the EDTA-gel samples.

Even so, the PM_S900 sample, whose mean d_g (8 ± 6 μm) is close to the values measured for EDTA_S700 (7 ± 4 μm), presents a value of ε_s unexpectedly high, which can be justified by the effect caused by the high dispersion of the grains size, producing significant changes on the orientation of the electric field lines in the interior of the sample.

In table 9, comparing the measurements performed at T= 360 K, it was found that samples produced by the PM method present lower corrected resistance for the grain boundaries ($R_{gb}.A/L$) than the samples produced by the EDTA-gel method. The brick layer model [8] predicts that microstructures with higher mean grain size should present higher macroscopic conductivity, i.e., lower $R_{gb}.A/L$.

From the analysis of the data it was concluded that the relation between the mean grain size (d_g) and $R_{gb}.A/L$ does not always follow the used model. Comparing the values of $R_{gb}.A/L$ among the PM samples we find that PM_S700 presents the major d_g value (84 ± 51 μm) and the lower $R_{gb}.A/L$ value, which is in accordance with the brick layer model. However, comparing PM_S800 and PM_S900 samples, we observed a divergence from the predictions of the model, once that although PM_S800 presents a higher d_g value (36 ± 22 μm) than the PM_S900 sample (8 ± 6 μm), the value of $R_{gb}.A/L$ measured for PM_900 sample is lower than the one obtained for PM_S800. PM_S900 sample is characterized by its heterogeneous microstructure, with a large dispersion of the grain size distribution. It is probable that the electric current tends to pass through the larger grains, decreasing, by this way, the number of the total crossed boundaries, which leads to the decrease of macroscopic resistance of the material. This kind of divergence was already predicted by Fleig and Maier [fleig99], as it was described in page 9.

3.6. Conclusion

After the presentation and analysis of the experimental data obtained for the ceramic bulk samples produced by the EDTA-gel and PM methods, it would be useful to highlight the most important results.

- At the microstructural level it can be pointed that, in general, samples produced by the PM method present a larger average grain size than the ones produced by the EDTA-gel method, although the applied calcination and sintering processes were the same. It is not clear a correlation between the sizes of the calcined powders and the average grain size of the pellets produced from those powders.
- Dielectric measurements show, for all the samples, the existence of two relaxation processes. One in the low frequency region, associated to the interfacial polarization process (sample surface–electrode) and the other in the MHz region and is usually associated to a Maxwell-Wagner relaxation type, resulting from the polarization at the interface grain–grain boundaries.
- The dielectric strength ($\Delta\varepsilon$) results obtained for the samples produced by these sol-gel methods (EDTA-gel and PM) are higher than those obtained for the samples produced by solid state reaction. Among the tested sol-gel methods, higher $\Delta\varepsilon$ results are obtained for the PM samples, reaching a maximum value ($\Delta\varepsilon = 13722$) for the PM_S700 sample. These results can be partially justified by the larger average grain size measured for the PM samples, which is in accordance with the predictions of brick layer model.
- The larger conductivity measured for the sample produced by sol-gel methods justify the larger tan δ of these samples compared with other reported measurements obtained for CCTO samples produced by solid state reaction.

REFERENCES

[1] Adams, T. B., Sinclair, D. C. & West, A. R. (2006). *Phys. Rev. B*, *73*, 094124.

[2] Almeida, A. F. L., Fechine, P. B. A., Góes, J. C., Valente, M. A., Miranda, M. A. R. & Sombra, A. S. B. (2004). *Materials Science and Engineering B – Solid State Materials for Advanced Technology*, 111, 113.

[3] Amaral, F., Rubinger, C. P. L., Henry, F., Costa, L. C., Valente, M. A. & Barros-Timmons, A. (2008). *J. Non Cryst. Sol.*, *354*, 5321.

[4] Amaral, F., Costa, L. C., Valente, M. A. & Henry, F. (2009). *J. Non Cryst. Sol.*, *355*, 2160.

[5] Amaral, F., Valente, M. A. & Costa, L. C. (2010). *Mat. Chem. Phys.*, *124*, 580.

[6] Amaral, F., Valente, M. A. & Costa, L. C. (2011). *Phys. Status Solidi A*, *124*, 2284.

[7] Bauerle, J. E. (1969). *J. Phys. Chem. Solids*, *30* (1969), 2657.

[8] Beeckmans, N. M. & Heyne, L. (1976). *Electrochimica. Acta*, *21*, 303.

[9] Brinker, C. J. & Scherer, G. W. (1990). *The Physics and Chemistry of Sol-Gel Processing*, Academic Press, New York.

[10] Chung, S. -Y., Choi, S. -Y., Yamamoto, T., Ikuhara, Y. & Kang, S. -J. (2006). *App. Phys. Lett.*, *88*, 091917

[11] Cole, K. S. & Cole, R. H. (1941). *J. Chem. Phys.*, *9* (1941), p. 341.

[12] Costa, L. C. & Mendiratta, S. K. (1994). *J. Non-Cryst. Sol.*, *172*, 324.

[13] van Dijk, T. & Burggraaf, A. J. (1981). *Phys. Stat. Sol. (a)*, 63, 229.

[14] Fang, L., Shen, M. & Cao, W. (2004). *J. Appl. Phys.*, *95*, 6483.

[15] Fang, L., Shen, M., Yang, J. & Li, Z. (2006). *Sol. St. Comm.*, *137*, 381.

[16] Feng, L., Tang, X., Yan, Y., Chen, X., Jiao, Z. & Cao, G. (2006). *Phys. Stat. Sol. (a)*, *203*, 1862.

[17] Fleig, J. & Maier, J. (1997). *J. Electroceramics*, *1*, 73.

[18] Fleig, J. & Maier, J. (1998). *J. Electrochem. Soc.*, *145* (1998), 2081.

[19] Fleig, J. & Maier, J. (1999). *J. Am. Ceram. Soc.*, *82*, 3485.

[20] Fleig, J. & Maier, J. (1999). *J. Eur. Ceram. Soc.*, *19*, 693.

[21] Fransaer, J., Roos, J. R., Delaey, L., Van Der Biest, O., Arkens, O. & Celis, J. P. (1989). *J. Appl. Phys.*, *65*, 3277.

[22] Fritsch, S. G., Nava, Z. V., Tenailleau, C., Lebey, T., Durand, B. & Chane-Ching, J. (2008). *Adv. Mater.*, *20*, 551.

[23] Hodgson, S. N., Shen, X. & Sale, F. R. (2000). *J. Mater. Sci.*, *35*, 5275.

[24] Jianren, L., Cho, K., Wu, N. & Ignatiev, A. (2004). *IEEE Trans. Diel. Elect. Insul.*, *11*, 534.

[25] Jin, S., Xia, H., Zhang, Y., Guo, J. & Xu, J. (2007). *Mater. Lett.*, *61*, 1404.

[26] Jonscher, A. K. (1983). *Dielectric relaxation in solids*. London: Chelsea Dielectric Press: London.

[27] Kahn, M. (accessed 15.02.13). Multilayer Ceramic Capacitors: Material and Manufacture, AVX technical Information, Myrtle Beach, SC, 2011. Available from: /http://www.avx.com/docs/techinfo/mlcmat.pdf.

[28] Laugier, J. & Filhol, A. (1978). *CELREF Program*, ILL (Grenoble, France).

[29] Li, J., Sleight, A. W. & Subramanian, M. A. (2005). *Solid State Communications*, *135*, 260.

[30] Liu, J., Duan, C. G., Mei, W. N., Smith, R. W. & Hardy, J. R. (2005). *J. Appl. Phys.*, *98*, 093703.

[31] Liu, J., Sui, Y., Duan, C., Mei, W. N., Smith, R. W. & Hardy, J. R. (2006). *Chem. Mater.*, *18*, 3878.

[32] Lu, J. & Wong, C. P. (2008). *IEEE Transactions on Dielectrics and Electrical Insulation*, *15*, 1322.

[33] Ma, Y., Liu, J., Gao, C., Mei, W. N., White, A. D. & Rasty, J. (). *Appl. Phys. Letters*, *88*, 191903.

[34] Martell, A. E. & Smith, R. M. (1982). *Critical Stability Constants*, vol. 5, Plenum Press, New York.

[35] Masingboon, C., Thongbai, P., Maensiri, S., Yamwong, T. & Seraphin, S. (2008). *Mater. Chem. Phys.*, *109*, 262.

[36] Maxwell, J. C. (1881). *A treatise on Electricity and Magnetism*, 2ª edição, Clarendon Press, Oxford.

[37] Metlin, Y. G. & Tretyakov, Y. D. (1994). *J. Mater. Chem.*, *11*, 1659.

[38] Moulson, A. J. & Herbert, J. M. (2003). *Electroceramics: Materials, Properties, Applications*, Jonh Wiley & Sons, New York.

[39] Nishino, A. (1996). *J. Power Sources*, *60*, 137.

[40] Pechini, M. P. (1967). U.S. Patent No. 3,330,697.

[41] Alam, M. A., Zuga, L. & Pecht, M. G. (2012). *Ceramics International*, *38*, 6091.

[42] Robertson, J. (2004). *Eur. Phys. J. Appl. Phys.*, *28*, 265.

[43] Azaroff, L. V. (1968). *Elements of X-ray Crystallography*, McGraw-Hill, New York.

[44] Shao, F., Zhang, J. L., Zhong, W. L. & Wang, C. L. (2006). *J. Appl. Phys.*, *99*, 084106.

[45] Sinclair, D. C., Adams, T. B., Morrison, F. D. & West, A. R. (2002). *Appl. Phys. Lett.*, *80*, 2153.

[46] Smith, R. M. & Martell, A. E. (1989). *Critical Stability Constants*, vol. 6, Plenum Press, New York.

[47] Stojanovic, B. D. (2005). *Ferroelectrics*, *319*, 65.

[48] Subramanian, M. A., Li, D., Duan, N., Reisner, B. A. & Sleight, A.W. (2000). *J. Sol. State Chem.*, *151*, 323.

[49] Subramanian, M. A. & Sleight, A. W. (2002). *Sol. Stat. Sci.*, *4*, 347.

[50] Vogel, A. I. (1989). *Vogel's textbook of quantitative chemical analysis*, Longman Scientific & Technical, Essex.

[51] Wang, H. W., Hall, D. A. & Sale, F. R. (1992). *J. Am. Ceram. Soc.*, *75*, 253.

[52] Electrical Industry Alliance (EIA), (2012). World Capacitor Trade Statistics (WCTA).

[53] Wilk, G. D., Wallace, R. M. & Anthony, J. M. (2001). *J. Appl. Physics*, *89*, 5243.

[54] Xu, Y. B., Huang, G. H. & Long, H. (2003). *Mater. Lett.*, *57*, 3570.

[55] Almeida, A. F. L., Oliveira, R. E. S., Sasaki, J. M., Sombra, A. S. B. & Kretly, L. C. (2003). *Micr. Opt. Technol. Lett.*, *39*, 145.

[56] Amaral, F., Rubinger, C. P. L., Valente, M. A., Costa, L. C. & Moreira, R. L. (2009). *J. Applied Physics*, *105*, 034109.

[57] Choy, J. & Han, Y. (1997). *J. Mater. Chem.*, *7*, 1815.

[58] Deng, G., He, Z. & Muralt, P. (2009). *J. Appl. Phys.*, *105*, 084106.

[59] Ferrarelli, M. C., Sinclair, D. C., West, A. R., Dabkowska, H. A., Dabkowski, A. & Luke, G. M. (2009). *J. Mater. Chem.*, *19*, 5916.

[60] He, L., Neaton, J. B., Cohen, M. H. & Vanderbilt, D. (2002). *Phys. Rev. B*, *65*, 214112.

[61] Kristine, F. J., Shepherd, R. E. & Siddiqui, S. (1981). *Inorg. Chem.*, *20*, 2571.

[62] Krohns, S., Lunkenheimer, P., Ebbinghause, S. G. & Loidl, A. (2008). *J. App. Phys.*, *103*, 084107.

[63] Macdonald, J. R. & Barsoukov, E. (2005). *Impedance Spectroscopy: Theory, Experiment and Applications*, Wiley, New York.

[64] Pineda, M., Fierro, J. L. G., Palacios, J. M., Silleruelo, C., Garcia, E. & Ibarra, J. V. (1997). *Appl. Surf. Sci.*, *119*, 1.

[65] Xu, J. F., Ji, W., Shen, Z. X., Li, W. S., Tang, S. H., Ye, X. R., Jia, D. Z. & Xin, X. Q. (1999). *J. Raman Spectrosc.*, *30*, 413.

In: EDTA
Editor: Andris Molnar

ISBN: 978-1-62808-146-6

Chapter 3

PHOTO- OR RADIATION-INDUCED DEGRADATION OF EDTA AND ITS COMPLEXES OF HEAVY METALS

Katerina Cubova and Barbora Drtinova*

Dept. of Nuclear Chemistry, Faculty of Nuclear Science and Physical Engineering, Czech Technical University, Brehova, Prague, Czech Republic

ABSTRACT

EDTA is present in a variety of waste, of which one of the most dangerous is liquid radioactive waste produced for years in nuclear power generation, defense applications, and various research institutions or medical applications of radionuclides. Some of these wastes (e.g. decontamination solutions or evaporator concentrates) contain, in addition to the radionuclides, also strong organic complexants like EDTA, generally used as decontamination and washing solutions. The presence of the organic complexants can cause gas generation, microbiological activity, void formation, elevated leachability and simultaneously higher mobility of cationic contaminants and thus can negatively influence the quality of the final forms of treated waste.

For the treatment of organic-bearing aqueous waste, photocatalytic degradation by 254 nm UV radiation with a TiO_2 photocatalyst can be

* E-mail: barbora.drtinova@fjfi.cvut.cz.

used. Systematic study of the decomposition of EDTA and its metal complexes has been carried out. It was found that while pure EDTA can be degraded with high efficiency, the decomposition of complexes is much more complicated. Separation of the radionuclides by ion exchange can follow the photocatalytic degradation of organic complexes and complexants in a multistage process developed and tested for the treatment of this type of wastes.
Also the utilization of ionizing radiation (by accelerated electrons or γ-irradiation) could lead to degradation of organic complexants and consequently to minimization of the volume and solubility of waste containing metal complexes with strong organic agents. E.g. EDTA and citric acid are radiolytically destroyed by •OH radicals with high rate constants achieving values about 10^9 $dm^3.mol^{-1}.s^{-1}$. Therefore the complexing agent could serve as an •OH radical scavenger enabling the radiation reduction of metal ions released after the destruction of complexes. Another possible mechanism is the precipitate formation consisting of products of organic complexant degradation and metal.

1. INTRODUCTION

Dangerous liquid radioactive wastes have been produced for years in nuclear power generation and defense applications as in various research institutions or medical applications of radionuclides. Some of these wastes (e.g. decontamination solutions or evaporator concentrates) contain, in addition to the radionuclides, many other substances such as heavy or toxic metals, often bound with organic complexing agents like EDTA. EDTA is one of the most common species that cause difficulties in the process of treatment of liquid radioactive waste containing complexing agents and thus can negatively influence the quality of the final forms of treated wastes. Because of the chemical structure the decomposition of complexing agents and separation of the radionuclides and heavy or toxic metals ions could represent rather complicated problem.

For the treatment of the liquid radioactive waste containing organic complexing or chelating agents such as oxalic and citric acids, EDTA and waste containing metal ions and metal complexes, only combined processes in more stages seem to be efficient. A method comprising one-stage photo degradation of organic complexants or complexes and the sorption of the released metal ions onto absorber was shown to be promising solution (Danacikova 1997, Danacikova 1998, Advances… 1994).

Photochemical mineralization of organic materials by UV radiation was demonstrated to be a prospective method for waste treatment (Legrini 1993, Litter 1994, Sabin 1992, Low 1991, Danacikova-Popelova 1999). As the method for the decomposition of EDTA a heterogeneously catalyzed photodegradation was used (Legrini 1993, Hoffmann 1995, Schiavello(ed.) 1997). The photodegradation efficiency can be enhanced by heterogeneously dispersed semiconductor photocatalyst (Legrini 1993, Hoffmann 1995, Tanaka 1989, Schiavello(ed.) 1997). For the optimization of the waste treatment, detailed studies aimed at the investigation of the conditions and mechanism of their photodegradation were carried out.

Similarly promising is the utilization of a radiation processing of wastewater and sludge containing toxic metals (including the spent decontamination solutions), which may be also complexed with organic complexants as EDTA or citric acid (Cubova 2012). The complexant present in solution can actually fulfill the function of an $^{\bullet}OH$ radical scavenger as can be seen in the case of lead complexed with EDTA (Drtinova 2006, 2008). Moreover, the ionizing radiation, due to its penetrating ability, is applicable even for turbid solutions.

2. HETEROGENEOUS PHOTOCATALYSIS

The term of heterogeneous photocatalysis is not easy to define; the mechanism of ongoing reactions is complicated and in many cases uncertain in details. In general, heterogeneous photocatalysis can be understood as a catalytic process involving oxidation and reduction reactions which are initiated by means of electron-hole pairs photogenerated on the surface of semiconductor materials illuminated by the light of suitable energy. The rates of photocatalytic reactions increase as a consequence of an optical excitation of a solid that remains chemically unchanged (Schiavello(ed.) 1997, Cernochova 2002).

The impacts of photons of suitable energy stimulate the charge separation. Electron-hole pairs formed in this way can subsequently evolve in different ways: they can recombine with emission of thermal energy and/or luminescence or they react with electron acceptor or donor species giving rise to reduction or oxidation processes, respectively. Upon illumination of suitable polycrystalline semiconductor nonhomogeneously suspended in aqueous or gaseous mixture the potential energy of photogenerated electrons and holes can be transferred to redox couples in the solution; electrons can reduce the oxidised species of the couple and holes can oxidise the reduced species of the couple. The ratio of the rates of redox processes to the rates of recombination determines the photocatalytic activity of the semiconductor (Schiavello(ed.) 1997, Cernochova 2002).

Titanium dioxide TiO_2 displays high photoactivity, adequate radiation and chemical stability and no toxicity. In addition, titanium dioxide is common, easy available and rather cheap compound widely used in industry (Cernochova 2002). Therefore polycrystalline TiO_2 is accepted as one of the most suitable semiconductor material for the photocatalytic degradation of organic species in wastewater (Schiavello(ed.) 1997).

It can be said that heterogeneous photocatalysis seem to be very valuable tool of degradation of organic pollutants. The current research interest is concentrated on the photocatalytic degradation of waste streams containing organic compounds in systems with semiconductor photocatalysts. Irradiation of the polycrystalline semiconductor dispersed in a liquid waste starts a complicated sequence of chemical reactions in which organic species present in the waste are partly or fully decomposed. The main interest is focussed on waste streams containing aromatic compounds such as phenol and chlorophenol, chelating materials such as EDTA and other carboxylic acids, waste containing metal ions and metal complexes (Pope 1998, Cernochova 2002). The main concern of this chapter is focused on the photodegradation of EDTA and its complexes with metals.

2.1. Photocatalytic Degradation of EDTA

To be able to understand and to optimize the conditions of photocatalytic degradation of EDTA a detailed mechanistic study was performed. The results revealed that EDTA could be effectively degraded in a small immersion well photochemical reactor of 80 mL capacity with 6W low-pressure mercury lamp, using TiO_2 photocatalyst. The photodegradation experiments were performed using the following standard method: 80 mL of the solution was poured into photochemical reactor and titanium dioxide photocatalyst was added. The reaction mixture was irradiated under continued bubbling of air at controlled temperature for a required period. Before the analysis, the titanium dioxide was separated from the suspension.

Two types of blank experiments were performed. In the first one [BE1], degradation of the EDTA was followed in absence of photocatalyst. The second one [BE2] tested the decrease of the concentration of EDTA without the influence of irradiation.

EDTA determination was performed by a spectrophotometric method adapted from the original method described by Kaiser (1973): 20 mL of the solution was mixed with 1 mL of 0.1 $mol.L^{-1}$ $CoCl_2$, 1 mL of 3 $mol.L^{-1}$ $NaNO_2$, 1 mL of 0.1 $mol.L^{-1}$ HCl and the mixture was stirred for 3 min. Then 2 mL of H_2O_2 were added and stirred again for 2 min. The extinction of the EDTA-Co complex was measured at 555 nm. In the study of EDTA-Co and EDTA-Cr complexes, Co and Cr were determined by atomic absorption spectrometry; the degree of decomposition of EDTA was determined by TOC (total organic carbon) analysis.

From the experiments several following conclusions can be drawn:

In agreement with the general literature data (Schiavello (ed.) 1997), the kinetics of the EDTA photodegradation was confirmed to be of the first order.

The rate of the photodegradation was found to be temperature-dependent. The experimental data obtained allowed calculation of the apparent activation energy E_A of the photomineralization process. Its value for EDTA was found to be as low as $E_A = 14.8$ $kJ.mol^{-1}$. This value unambiguously indicates that the rate-determining step of the process is transport phenomena.

Experiments carried out to establish the influence of different amount of the photocatalyst revealed that the proportion of EDTA degraded increases rapidly with the increasing amount of TiO_2 in the reaction mixture. Blank experiments performed showed that significant part of this degradation is only apparent. The decrease of EDTA concentration in the solution was namely ascribed to its sorption onto the surface of the photocatalyst.

The experiments performed in systems saturated with nitrogen instead of air confirmed the important role of oxygen in the photodegradation processes. In the absence of oxygen, the degradation rate of EDTA has not exceeded one quarter of that achieved in oxygen saturated systems.

The effect of pH on the photocatalytic activity of TiO_2 was studied in the range of pH from 3 to 11 (Figure 1) because some literature data (Malinka 1994) indicated rather strong dependence of the photocatalytic activity of the TiO_2 catalysts on pH. The value of the blank experiment performed under irradiation but without addition of titanium dioxide (BE1) was found to increase significantly with increasing pH. These results could have been explained by correlating the blank experiment values with the dependence of the abundance of EDTA species on pH (calculated using the MINEQL code (Visual Minteq 1999)) (Figure 1). It was concluded that the dissociated ions $HEDTA^{3-}$, $EDTA^{4-}$ and $NaEDTA^{3-}$ are highly photolabile, while the lower charged EDTA species are, in absence of titanium dioxide, rather photoresistant.

The marked decrease of net EDTA degradation was observed at pH ≈ 5. This value corresponds well to the value of isoelectric point of the TiO_2 photocatalyst. The proposed mechanism for the decrease of the activity of the catalyst above the iso-pH is that at such values of pH the surface of the photocatalyst bears a negative charge and hence cannot sorb the negatively charged ions of dissociated EDTA. The degradation of EDTA by means of direct oxidation of the substrate sorbed onto the surface of the catalyst, which is one of the main degradation mechanisms (Schiavello(ed.) 1997, Rosikova 1998, Pope 1998, Danacikova 1995) is then not possible.

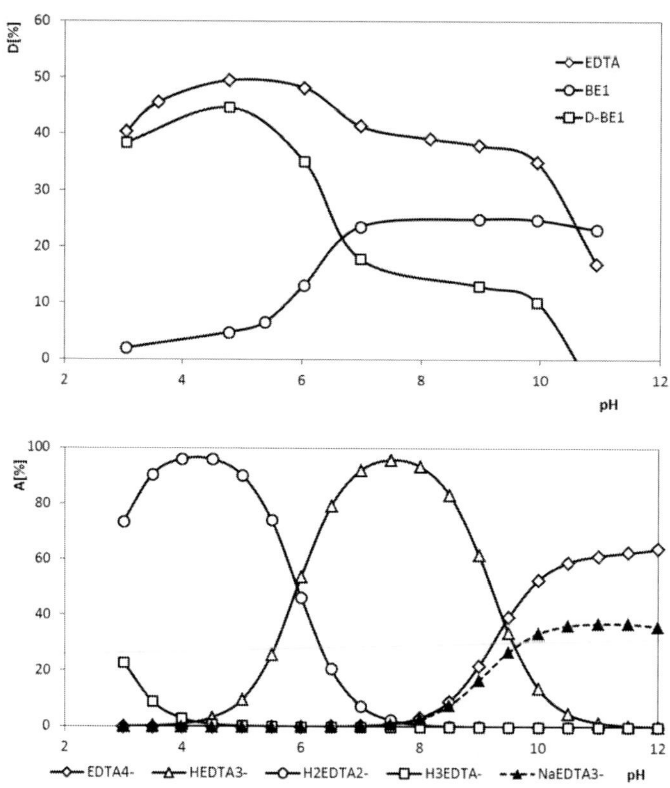

Figure 1. Degradation (D) of EDTA, and the value of blank experiment (BE_1, decomposition in absence of TiO_2) as a function of pH and relative abundance (A) of EDTA species as a function of pH (0.005 $mol.L^{-1}$ EDTA, t_{irrad} = 2 hrs, 1.25 $g.L^{-1}$ TiO_2 T = 25 ^0C, v_{air} = 24 $L.hr^{-1}$).

The photocatalytic activity of TiO_2 catalysts of different origin is known to differ significantly (Rosikova 1999, Danacikova-Popelova 1999, John 1997, Cernochova 2002, Danacikova 1999). Therefore the photocatalytic activities of five different photocatalysts were compared. The rate of EDTA photodegradation was studied (Rosikova 1999, Danacikova-Popelova 1999, Popelova 2002, Rosikova 1998) at the natural pH of the solution used (pH = 4.7) and at pH = 7 and 10. Characteristics of the TiO_2 catalysts used are introduced in Tab. 1. From the data, it was concluded that not only the photocatalytic activity of the catalysts varies with pH, but also that the behaviour of the single photocatalysts differs significantly (Černochova 2002, Rosikova 1999). The net photocatalytic activity is obtained by correlating the experimental results for both the blank experiments. Comparison of the photocatalysts based on the correlation of these data is shown in Figure 2. The results show that, the "net" – really induced by heterogeneous photocatalysis – degradation of EDTA is different for various catalysts.

Table 1. Characteristics of the TiO_2 photocatalysts used (Popelova 2002)

Catalyst	Specific surface area [$m^2\ g^{-1}$]	Crystallinity	Particle size	Admixtures
Degussa P 25	49.4	70 % anatase, 30 % rutile	~ 0.5 μm	N/A
TiO_2 – M	247.3	100 % anatase	1 μm	Al, S
HTO	174.0*	amorphous	N/A	N/A
AV-01	8.6	100 % anatase	~ 1μm	Al,K
RG-14	18.3	100 % rutile	~ 1μm	Al,Si

* Determined for vacuum dried (at room temperature) sample crushed approximately to the original grain size.

The difference among the photoactivities cannot be simply correlated with the specific areas of the catalysts and nor they seem to be connected with their crystalline structure. The photoactivity of the catalyst AV-01 (anatase) is nearly negligible while the photoactivity of TiO_2-M (anatase as well) was fairly high and Degussa P-25 (mixture of anatase and rutile) showed the highest activity in all the range studied. The activity of the photocatalysts is obviously determined by other physico-chemical properties not determined in this study.

Among the potential candidates, the surface density of OH-groups, width of the band gap, porosity, etc. should be listed. In the tests no simple correlation of the photocatalytic activity of TiO_2 materials and their specific surface areas was found – the specific surface area is not directly proportional to the activity of the photocatalysts. It seemed that much more significant influence on the efficiency of the photocatalyst has the method of preparation of the respective titanium dioxide.

2.2. Photocatalytic Degradation of Metal Complexes with EDTA

The previous studies (Rosikova 1999, Danacikova-Popelova 1999, John 1997, Popelova 2002, Rosikova 1998) demonstrated that the EDTA, oxalic and citric acids could be degraded with a high yield in a photocatalytic process.

The photodegradation of metal complexes of the mentioned acids may, however, differ significantly. Therefore the photocatalytic degradation of metal complexes with EDTA was performed. The choice of macrocomponents was given by the composition of the steel used for the equipment in primary circuit of VVER (water-cooled water-moderated power reactor) type in nuclear power plants because some of the complexants are used as the decontamination solution in NPPs.

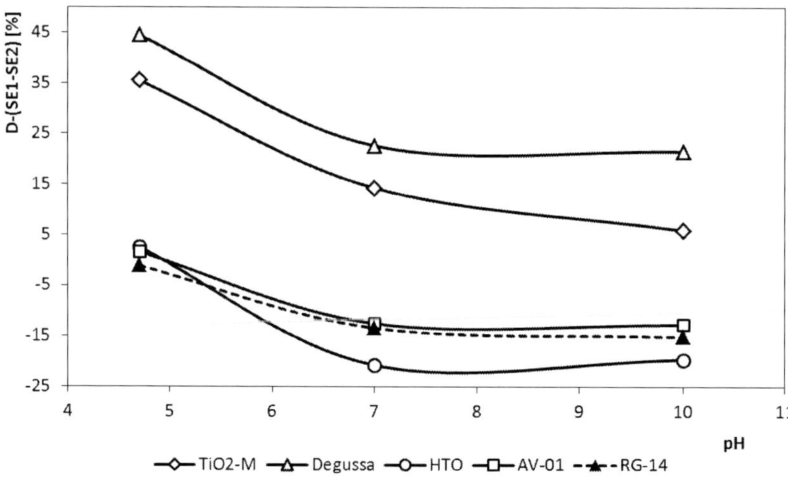

Figure 2. Net photocatalytic degradation of EDTA after substraction of blank experiments [D-(SE1+SE2)] in the presence of various photocatalysts at various pH values (0.005 mol.L^{-1} EDTA , t_{irrad} = 1 hr, 1.25 g.L^{-1} TiO_2, T = 25 ^{0}C, v_{air} = 20 L.hr^{-1}).

As a representative of the macrocomponents iron and chromium were selected. For the information on the behaviour of trace elements, also cobalt was included into the study.

2.2.1. Cobalt (Co^{2+}) Complexes with EDTA

Photocatalytic degradation of complexes of EDTA with Co(II) was studied for EDTA and $CoCl_2$ concentration 0.005 mol.L^{-1} in the pH range pH ~ 3 - 11. The degradation of organic compounds in this system was found to be very low (4 - 10 % over 1 hr of illumination). Regardless of the pH of the solutions studied, the concentration of cobalt decreased by approximately 12 - 15 % during all the photodegradation experiments.

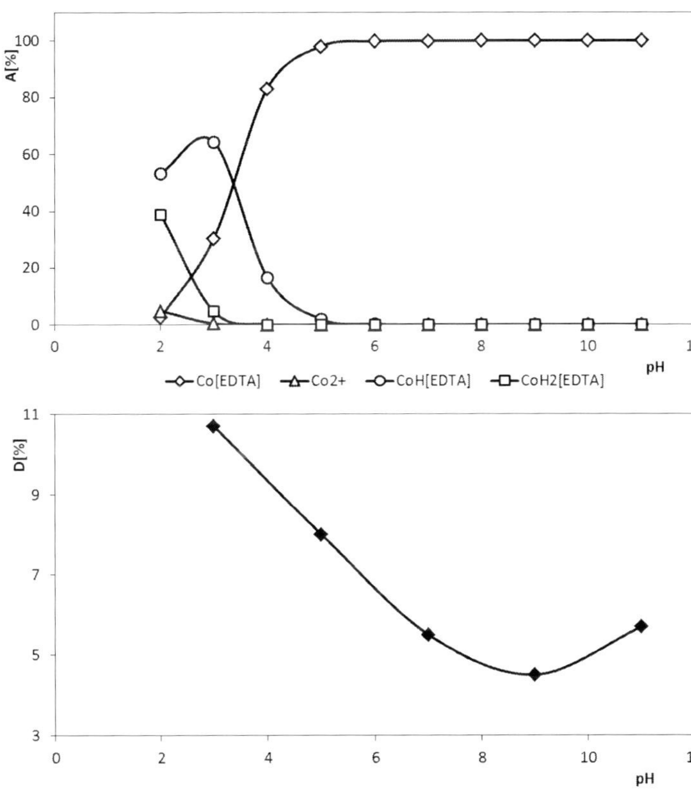

Figure 3 Degradation (D) of EDTA - Co complexes as a function of pH and the abundance (A) of EDTA - Co species on pH (calculated by MINTEQ) - 0.005 mol.L^{-1} EDTA + 0.005 mol.L^{-1} $Co(NO_3)_2$, t_{irrad} = 2 hrs, 1.25 g.L^{-1} TiO_2-M, t = 24 ^{0}C, v_{air} =26 L.hr^{-1} ; species with abundance > 5 % are presented, only.

Detailed explanation of the behaviour of EDTA - Co complexes can be done from the comparison of its degradation with the abundance of EDTA - Co species (calculated using the MINTEQ (Visual Minteq 1999)) (Figure 3). From this comparison it can be seen that the complex $CoEDTA^{2-}$ is formed (100 %) already at pH ~ 5 and is the only significant species up to pH = 13. This is probably also the reason why the solution is highly photostable and practically undegradable by the UV radiation. A small degree of degradation which was observed mainly in acidic range of pH can be ascribed to the direct degradation of the molecules of negatively charged complexes $CoHEDTA^-$ or $CoEDTA^{2-}$ sorbed on the TiO_2 surface. With increasing pH the charge of the TiO_2 surface switches to negative and the sorption, and hence also the degradation, decrease.

2.2.2. Chromium (Cr^{3+}) Complexes with EDTA

Photocatalytic degradation of complexes of EDTA with Cr^{3+} was performed in excess of EDTA (the solution of 0.0075 $mol.L^{-1}$ EDTA and 0.00125 $mol.L^{-1}$ $CrCl_3$ was used). The pH range tested was pH ~ 3 - 11. The results obtained were similar to those obtained for the EDTA - Co complexes. The degradation of EDTA - Cr complex was generally low (2 - 8 %) and decreased with increasing pH. When comparing the results obtained with the calculated abundance of the individual species in this system, it can be seen that the degradation observed does not correlate with the abundance of EDTA - Co complexes - chromium is 100 % complexed in all the pH range (in $CrEDTA^0$ complex from pH = 3 to pH ~ 7.5, and in $CrOHEDTA^{2-}$ complex at pH ≥ 7.5). The degradation does nor correlates with the abundance of the non - complexed EDTA species, either.

From the results described it can be concluded that both of the complexes (EDTA - Co and EDTA - Cr) are rather photostable and can not be degraded by direct action of UV radiation. Their degradation, if any, probably proceeds only by the mechanism of direct photooxidation or photoreduction of the molecules sorbed on TiO_2 surface by the electron - hole pairs.

2.2.3. Comparison of the Degradation of EDTA with EDTA - Co or EDTA - Cr Complexes

The experiments with metal complexes (Co, Cr) of EDTA showed that while pure EDTA can be degraded with high efficiency, the decomposition of its complexes is much more complicated.

Another difference is that the degradation of the EDTA - Co complex decreases with increasing pH. A potential explanation of these effects may be connected with chemical and photo-stability of the respective complexes. From the correlation of EDTA degradation with the abundance of EDTA species in pure EDTA, EDTA - Co, and EDTA - Cr systems (Figure 1 and 4), several interesting conclusions follow:

- EDTA – Co complexes seem to be rather photoresistant, regardless of the type of the complex (CoH_2EDTA^0, $CoHEDTA^-$ or $CoEDTA^{2-}$). Any significant photodegradation proceeds on the surface of the photocatalyst, only.
- EDTA – Cr complexes seem to be even more photoresistant, they cannot degrade on the surface of the photocatalyst, since no sorbable species exist in any pH range.
- H_3EDTA^- and H_2EDTA^{2-} ions are rather resistant to direct photolysis in absence of the TiO_2 photocatalyst. H_2EDTA^{2-} complex readily degrades when sorbed onto the (positively charged) surface of photocatalyst, while the H_3EDTA^- complex seems to be resistant even to degradation on the surface.
- $EDTA^{3-}$ and $EDTA^{4-}$ can be efficiently degraded even by direct photolysis in absence of a photocatalyst. These ions occur in the solution at pH ≥ 6, when the surface of the TiO_2 photocatalyst bears a negative charge and, hence, they can not sorb onto it. Therefore their photodegradation is not much more efficient in a system catalysed by TiO_2.
- the presence of Co^{3+} ions seems to have a strong inhibitive influence on the photocatalytic activity of TiO_2 since in its presence even the free, non - complexed, EDTA species are resistant towards photodegradation (the concentration of free non - complexed EDTA in the EDTA - Co system was approximately equal to the EDTA concentration in the experiments with pure EDTA).

Similar experiments were performed for oxalic and citric acids and their complexes with Co and Fe. Contrary to EDTA, the degradation of citrates and oxalates bound in metal complexes was found to be more efficient than the degradation of pure complexants. This finding was used for the proposal and development of the method for the treatment of the liquid wastes coming from the decontamination of primary circuit equipment in NPP (Cubova 2012, Rosikova 2003).

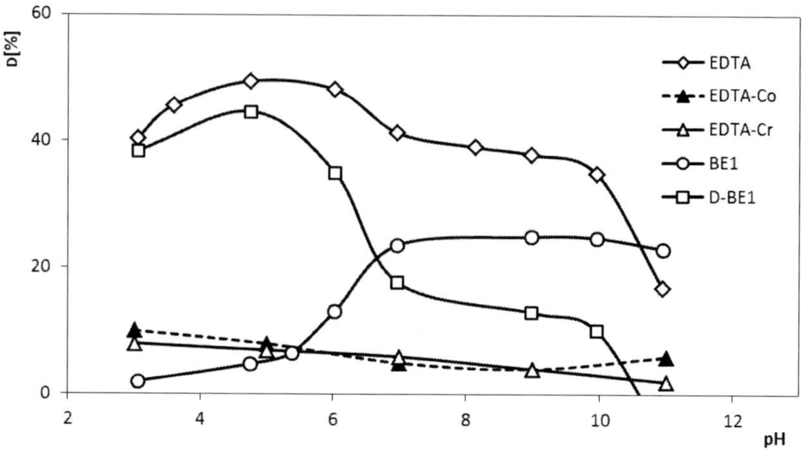

Figure 4. Degradation (D) of EDTA, complexes EDTA-Co, EDTA-Cr and the value of blank experiment (BE_1, decomposition in absence of TiO_2) as a function of pH (0.005 $mol.L^{-1}$ EDTA or 0.005 $mol.L^{-1}$ EDTA + 0.005 $mol.L^{-1}$ $Co(NO_3)_2$ or 0.005 $mol.L^{-1}$ EDTA + 0.005 $mol.L^{-1}$ $CrCl_3$, t_{irrad} = 2 hrs, 1.25 $g.L^{-1}$ TiO_2, T = 25 ^0C, v_{air} = 24 $L.hr^{-1}$).

3. Water Radiolysis

The general equation of water radiolysis, the process caused by the action of ionizing radiation (via γ-rays or using electron beam), is (12):

$$H_2O \rightarrow e^-_{aq}, H^\bullet, {}^\bullet OH, HO_2^\bullet, O^{\bullet\bullet}, H_2, H_2O_2, H_3O^+ (H^+_{aq}), OH^-_{aq} \quad (1)$$

During irradiation a great amount of energy is deposited in the aqueous solution where it induces a local temperature rise (by the absorption of energy of 1 MeV temperature rises up to 1010 K (Dolin 1973)). In the solution with the concentration of dissolved substances lower than 10 % (wt), practically all the energy is absorbed by water molecules (Swallow 1982) in the processes of excitation, ionization or dissociation leading to formation of highly unstable and reactive excited particles (1) which participate in reactions typical for radiation chemistry. The radiolysis of water can be divided into three stages (Bednar 1990) starting with the physical stage at the time t 10^{-16} s after the primary ionization at the level of the electron shells. At the time t 10^{-14} s in the physical-chemical stage the first dissociations occur. In the third, chemical stage beginning at t 10^{-11} s e^-_{aq}, ${}^\bullet OH$, H_3O^+ and other radicals and fragments

($HO_2^{\bullet}$ can be neglected for low linear energy transfer, $O^{\bullet\bullet}$ also occurs with very low yield) located in the spurs and blobs undergo reactions whose products and intermediates escape into the environment, being homogeneously distributed. This process and the third stage term at the time t 10^{-7} s.

Strongly reducing hydrated electrons e^-_{aq} (standard redox potential E^0 $(e^-_{aq}/e^-) = -2.9$ V), $H^{\bullet}$ radicals (E^0 $(H^+_{aq}/H^{\bullet}) = -2.3$ V) and strongly oxidizing $^{\bullet}OH$ radicals are the main species participating in reactions typical for the processes of radiation removal. Hydrated electrons e^-_{aq}, the thermalized electrons surrounded by oriented water molecules, are created with radiation-chemical yield of 0.28 μmol.J^{-1} (Buxton 1987). The radiation-chemical yield of $H^{\bullet}$ radical is 0.062 μmol.J^{-1} (Buxton 1987). Its reducing ability increases with increasing pH. Under strongly alkaline conditions, the hydrogen radical is subject to reaction (2) resulting in formation of hydrated electron.

$$H^{\bullet} + OH^-_{aq} \rightarrow e^-_{aq} + H_2O \qquad (2)$$

On the other hand, in very acidic environment it has the oxidizing properties. $^{\bullet}OH$ radicals are formed with the same radiation-chemical yield as hydrated electrons (Buxton 1987). Oxidative ability of $^{\bullet}OH$ radicals decreases with increasing pH.

For most applications is necessary to create only oxidizing or only reducing conditions while during radiolysis a comparable amount of strong oxidizing and strong reducing radicals is formed. Saturation of the solution with N_2O prior to irradiation leads to formation of oxidizing environment through the conversion of e^-_{aq} to $^{\bullet}OH$ radicals (reaction rate constant $k = 9.1\times10^9$ dm^3.mol^{-1}.s^{-1} at pH = 7) (3,4).

$$e_{aq}^- + N_2O \rightarrow N_2O^{\bullet -} \rightarrow N_2 + O^{\bullet -} \qquad (3)$$

$$N_2 + O^{\bullet -}(H_2O) \rightarrow {}^{\bullet}OH + OH^- + N_2 \qquad (4)$$

N_2O reacts very slowly also with $H^{\bullet}$ atoms ($k = 1.3\times10^4$ dm^3.mol^{-1}.s^{-1} in pH range 3.5 - 11). In oxidizing environment the radiation decomposition of organic substances such as EDTA and citric acid proceeds (Su 1998). On the contrary, the reducing environment is necessary for a successful radiation induced reduction of metal cations. Appropriate conditions are reached by adding of a scavenger of $^{\bullet}OH$ radicals (usually simple alcohols or formate anion) reacting according to (5) into irradiated system while mixture of organic radicals with reducing properties (weaker than e^-_{aq}) is formed.

$$HCOO^{-} + {}^{\bullet}OH \rightarrow COO^{\bullet -} + H_2O(k = 3.1\times10^{9}\ dm^{3}.mol^{-1}.s^{-1}) \quad (5)$$

It is also common to bubble the solution before irradiation with an inert gas to prevent reoxidizing of the products or the intermediates of the radiation reduction by diluted oxygen.

3.1. Radiation Degradation of EDTA

OH radicals are responsible for degradation of EDTA via oxidizing mechanism (Su 1998) as can be seen from the values of the rate constants (reaction 6-8) and radiation-chemical yields of water radiolysis products (Buxton 1987).

$$EDTA + H^{\bullet} \rightarrow (k = 6.5\times10^{7}\ dm^{3}.mol^{-1}.s^{-1}) \quad (6)$$

$$EDTA + e^{-}_{aq} \rightarrow (k=1\times10^{8}dm^{3}.mol^{-1}.s^{-1}) \quad (7)$$

$$EDTA + {}^{\bullet}OH \rightarrow (k=2.8\times10^{9}dm^{3}.mol^{-1}.s^{-1}) \quad (8)$$

By means of pulse radiolysis, the rate constants for reaction (9-11) of ${}^{\bullet}OH$ radicals with the different species of EDTA were determined (Höbel 1998).

$$OH + H_2EDTA^{2-} \rightarrow EDTA\ radicals\ (k=4\times10^{8}dm^{3}.mol^{-1}.s^{-1};pH4) \quad (9)$$

$$OH + HEDTA^{3-} \rightarrow EDTA\ radicals\ (k=2\times10^{9}dm^{3}.mol^{-1}.s^{-1};pH9) \quad (10)$$

$$OH + EDTA^{4-} \rightarrow EDTA\ radicals\ (k=6\times10^{9}dm^{3}.mol^{-1}.s^{-1}) \quad (11)$$

As in the case of heterogenous photocatalysis, the ionizing radiation can also produce pairs e^{-} - h^{+} in the semiconductor catalysts as TiO_2 and since it is more penetrating, the method is applicable as well to turbid solutions (Su 1998). (Another manifestation of the presence of the solid phase may be purely the adsorption effect, which relates more to the removal of metals.)

The effect of 1.25×10^{-3} mol.L^{-1} of O_2, 1.9 mg.l^{-1}.min^{-1} O_3, or addition of 2.5 mg.L^{-1} TiO_2 on the efficiency of photolysis and radiolysis in degradation of 1×10^{-3} mol.L^{-1} EDTA was compared (Krapfenbauer 1999, Getoff 2001). In the photochemical process, the molecules of EDTA are directly excited with

$\lambda = 254$ nm.O_3 is photodegradable, during the process $^{\bullet}OH$ radicals as its most important products are formed. After TiO_2 produces a pair of e^- - h^+, $^{\bullet}OH$ radicals are created in reaction (12).

$$TiO_2^+ + H_2O \rightarrow TiO_2 + {}^{\bullet}OH + H^+ \quad (12)$$

The photochemical degradation increases in order: bubbling with O_2; ozonation, and ozonation with TiO_2 addition (the complete degradation of EDTA was achieved in 30 minutes).

Table 2. Products of photochemical degradation of aqueous solution of EDTA (1×10^{-3} mol.L^{-1}) (GC/MS analysis, pH = 6.5, 0.25×10^{-3} mol.L^{-1} O_2, absorbed dose UV: I = 3.84×10^{18} hν.mL^{-1} at λ = 253.7 nm; E = 4.88 eV.hν^{-1}) (Krapfenbauer 1999)

Products of photochemical degradation	μmol.L^{-1}	Calculated G_i^{ph} (for dose of 1 kGy)
Degradation of EDTA	643	2.2
N-methyltriacetylethylendiamine	196	0.6
N-hydroxymethyliminodiacetic acid	543	1.7
Iminodiacetic acid	175	0.5
N-acetylethylendiamine	Not determined	

Table 3. Products of radiation degradation of aqueous solution of EDTA (1×10^{-3} mol.L^{-1}) (GC/MS analysis, pH = 6.5, 0.25×10^{-3} mol.L^{-1} O_2, absorbed dose 2kGy; dose rate 126 Gy.min^{-1}) (Krapfenbauer 1999)

Products of radiation degradation	μmol.L^{-1}	Calculated G (for dose of 2 kGy)
Degradation of EDTA	332	1.6
N,N´-methylacetylethylendiamine	260	1.2
N-hydroxymethyliminodiacetic acid	120	0.6
Iminodiacetic acid	60	0.3
N-acetylethylendiamine	stopy	
CO_2	294	2.3
NH_3	32	0.3

The activated TiO_2 with 0.1 % (wt) Co was used in the experiment studying the effect of γ-irradiation. Bubbling with O_2 was again the least efficient process, followed by comparative ozonation in the non-irradiated system, ozonation, bubbling with O_2 with addition of TiO_2, and ozonation with TiO_2 addition. 95 % degradation of EDTA was achieved at the dose of 1.5 kGy for the ozonation with the addition of TiO_2 and at 8 kGy for the bubbling with O_2. In the ozonated system containing TiO_2, photodegradation and radiolysis yields are comparable. Products of both processes differ slightly (see Table 2 and 3).

3.2. Radiation Removal of Selected Divalent Metals

It is generally assumed that the removal of metals such as Pb or Cd is in the presence of $^{\bullet}OH$ radicals scavengers carried out as a process of radiation reduction. This knowledge conforms to the behavior of Pb^{2+} ions dissolved from $PbCl_2$ (Chaychian 1998) which are reduced by hydrated electrons (13), it is followed by disproportionation of Pb^+ ions (14) and in parallel dimerisation probably runs. The final precipitation is the result of clustering of Pb atoms (15).

$$Pb^{2+}+e^-_{aq}\rightarrow Pb^+(3.9\times10^{10}\ dm^3.mol^{-1}.s^{-1}) \qquad (13)$$

$$2Pb^+ \rightarrow Pb + Pb^{2+}(1.7\times10^8 dm^3.mol^{-1}.s^{-1}) \qquad (14)$$

$$nPb \rightarrow Pb_n \qquad (15)$$

Hydrogen radicals are unable to react with the Pb^{2+} ions but react with the Pb^+ ions (16). PbH^+ then disproportionates again (17).

$$Pb^++H^{\bullet}\rightarrow PbH^+(3\times10^9\ dm^3.mol^{-1}.s^{-1}) \qquad (16)$$

$$2PbH^+\rightarrow Pb+Pb^{2+}+H_2(2\times10^8\ dm^3.mol^{-1}.s^{-1}) \qquad (17)$$

On the other hand, in the reaction with $^{\bullet}OH$ radicals Pb^{3+} ions are formed and their reaction with Pb^+ leads back to formation of Pb^{2+} ions. Since the reducing and oxidizing products of water radiolysis (1) are created in comparable amount, the presence of an $^{\bullet}OH$ radical scavenger is necessary for the radiation reduction of dissolved metal (Precipitation of Pb during irradiation of aerated or deoxygenated solutions does not take place) (Chaychian 1998, Zhestkova 1997).

One of the most widely used $^{\bullet}OH$ radical scavenger is formate anion (Zhestkova 1997). Ionradical $COO^{\bullet-}$, capable to reduce both the Pb^{2+} and Pb^{+} (20,21), is created in reactions (18,19).

$$HCOO^{-} + {}^{\bullet}OH \rightarrow COO^{\bullet-} + H_2O \quad (18)$$

$$HCOO^{-} + H^{\bullet} \rightarrow COO^{\bullet-} + H_2 \quad (19)$$

$$Pb^{2+} + COO^{\bullet-} \rightarrow Pb^{+} + CO_2 \quad (20)$$

$$Pb^{+} + COO^{\bullet-} \rightarrow Pb + CO_2 \quad (21)$$

The main advantages of utilization of formate are its strong behavior as a scavenger of $^{\bullet}OH$ radicals and the non-toxicity of the final product, CO_2 (20,21), while e.g. using ethanol as a scavenger results in the creation of toxic acetaldehyde (Clearfield 1982).

Aforementioned applies to the radiation reduction of metallic ions. However, it is not the only mechanism leading to the successful removal of metal from solution. 100 $mg.L^{-1}$ of Pb^{II} can be removed from aqueous solutions in the presence of HCOOK or isopropanol as $^{\bullet}OH$ radical scavengers by means of accelerated electrons (Pospisil 2006, Drtinova* 2010). In the system with 10% isopropanol, 95% removal of lead occurs at the absorbed dose of 30 kGy and the radiation reduction of lead occurs (Drtinova* 2010). On the other hand, the complete removal from aqueous solutions (pH ~ 5 – 6) containing 1×10^{-2} $mol.L^{-1}$ of HCOOK can be achieved already at the dose of 2.5 kGy. The product of dissolved Pb^{II} irradiation depends strongly on formate concentration and the salt from which is lead originally dissolved. It could vary from metallic lead to a precipitate of $PbCO_3$ as was proved by the thermal analysis. The radiation precipitation of $PbCO_3$ occurs as an increasing amount of $Pb(Formate)^{+}$ species is present in the solution prior to irradiation which increases with increasing concentration of HCOOK.

The behavior of dissolved cadmium in the irradiated solution is very similar to that of lead (reactions 13-15 take place). In contrast, the reactions of Cd^{2+}or Cd^{+} (16) with the $H^{\bullet}$ radicals do not run (Zhestkova 1997). For successfully removal of cadmium (100 $mg.L^{-1}$), the absorbed dose of 15 kGy is at least necessary (for the system with 1×10^{-2} $mol.L^{-1}$ of HCOOK). The products vary according to the type of action that takes place in the irradiated system. $CdCO_3$ is the product in the system with $Cd(NO_3)_2$ as the initial salt and Cd^0 in the system with cadmium dissolved from $CdCl_2$ (Drtinova 2010). It shows that present anion of starting salt significantly affects the behavior of dissolved metal under irradiation (see Table 4).

For both metals the combined effect of irradiation with adsorption on solid phase (bentonite, active carbon, zeolite, NiO, TiO_2, Cu_2O and CuO - contact time = 24 hrs) was studied (Pospisil 2006, Drtinova 2010).

The removal of metals from irradiated solutions was negatively influenced with the present solid phase (0.125 - 1 $mg.mL^{-1}$) with two exceptions - zeolite (Pospisil 2006) and Cu_2O.

Presence of Cu_2O very unexpectedly directly influences the behavior of irradiated system containing only dissolved salt of lead without any addition of $^{\bullet}OH$ radical scavenger. Approximately 50% removal of lead is achieved at a dose of 1.5 kGy and remains constant with increasing absorbed dose studied up to 20 kGy.

3.3. Radiation Removal of Comlexed Metals

EDTA as such has the properties of a quite strong scavenger of $^{\bullet}OH$ radicals (see 5 and 8-11). This feature can be successfully used for the combined radiation induced removal of metal complexd with EDTA (Chaychian 1998, Drtinova 2006,2008,2010).

The radiation removal of lead (1×10^{-3} $mol.L^{-1}$) dissolved from $PbCl_2$ complexed with EDTA (1.2×10^{-3} $mol.L^{-1}$) via accelerated electrons (50 Gy per pulse) was studied (Chaychian 1998). Practically all lead as well as EDTA is present as a species $(PbEDTA)^{2-}$ which undergoes reaction (22) leading to a partially reduced complex. After further reduction or disproportionation the lead precipitate is formed.

$$PbEDTA^{2-} + e^-_{aq} \rightarrow PbEDTA^{3-} (k = 1.9\times10^9\ dm^3.mol^{-1}.s^{-1} \text{ (Buitenhuis 1977)) } \quad (22)$$

97% removal of lead is achieved at an absorbed dose of 100 kGy at pH = 8 (adjusted by NaOH). At pH = 11 increased by adding of carbonate (4×10^{-3} $mol.L^{-1}$) as buffer which reduces the concentration of resulting H_3O^+ ions, the efficiency of radiation process achieved was the same. Carbonate acts also as a scavenger of $^{\bullet}OH$ radicals, while it reacts very slowly with solvated electrons (see Table 4).

Table 4. Rate constants of reactions of selected substances with primary radiolysis products (Bolton 1998)

Rate constants of reactions ($dm^3.mol^{-1}.s^{-1}$)			
	e^-_{aq}	$H^\bullet$	$^\bullet OH$
bicarbonate	1.0×10^6	4.0×10^4	8.5×10^6
carbonate	3.5×10^5	N/A	3.9×10^8
bromide	N/A	2.8×10^7	1.1×10^{10}
nitrate	9.7×10^9	1.4×10^6	N/A
nitrite	3.5×10^9	7.1×10^8	1.0×10^{10}
oxygen	1.9×10^{10}	2.1×10^{10}	N/A

The rate constants of reactions analogous to reaction (22) for some divalent metals increase with increasing ionic strength (Buitenhuis 1977). E.g. for cadmium when changing the ionic strength of solution from 0.016 (pH = 11.5) to 0.2 (pH = 12), the rate constant increases twice, from 2×10^8 to 3.9×10^8 $dm^3.mol^{-1}.s^{-1}$. For the lead at the same change of ionic strength the rate constant also increases twice, from 1.9×10^9 to 3.8×10^9 $dm^3.mol^{-1}.s^{-1}$.

Behavior of irradiated system containing complexed lead and cadmium (Drtinova 2006, 2008, 2010) was thoroughly studied. The water solutions of $Pb(NO_3)_2$, $PbCl_2$ and $Pb(CH_3COO)_2$ (4.8×10^{-4} $mol.L^{-1}$ = 100 $mg.L^{-1}$) mixed with 6×10^{-4} - 1×10^{-2} $mol.L^{-1}$ EDTA or $Cd(NO_3)_2$ and $CdCl_2$ (2.37×10^{-4} $mol.L^{-1}$ = 27 $mg.L^{-1}$) mixed with 3×10^{-4} $mol.L^{-1}$ EDTA (in detail in Drtinova 2013) were prepared. 1×10^{-3} - 1×10^{-2} $mol.L^{-1}$ of HCOOK, 2×10^{-3} $mol.L^{-1}$ of sodium carbonate CO_3^{2-}, 2×10^{-3} $mol.L^{-1}$ of sodium bicarbonate HCO_3^-, or 10% methanol were used as the $^\bullet OH$ radical scavengers. All the reagents were of analytical purity.

The addition of solids Cu_2O, Fe_2O_3, ZrO_2, W_2O_3 and TiO_2 (analytical grade) were also tested (contact time 24 hrs prior to irradiation). 8mL samples sealed in thin-glass ampoules were irradiated by accelerated electrons (4.5 MeV, dose rate of 1-1.5 $kGy.s^{-1}$) under vigorous stirring. Absorbed doses range from 0 up to 150 kGy as confirmed by means of alanine dosimeters. Subsequently, finely dispersed metallic particles and precipitated compounds were separated by centrifugation (5000 rpm). The residual concentration of metal ions in supernatant was determined by flame atomic absorption spectroscopy within an error of ± 3 %.

A “normalized concentration” in the radiation removal process may be described by (23)

$$N_C = [Me^{II}]/[Me^{II}]_0, \quad (23)$$

where $[Me^{II}]_0$ is the initial concentration and $[Me^{II}]$ is the post-irradiation concentration of soluble metallic species. The speciation modelling of solutions containing both metals and citric acid before irradiation was performed using PHREEQC program with MINTEQ database. Data for formic acid and its complexes with cadmium were reloaded from MINTEQ4 database (Visual Minteq 1999).

Lead in the system containing EDTA under conditions studied was present as a species $Pb(EDTA)^{2-}$ at pH > 4 independently of the anion of the initial salt (chloride, nitrate or acetate).The starting concentration used were half the values of concentration studied in (Chaychian 1998), the work on which the subsequent research was based.

The 92% lead ($Pb(NO_3)_2$) removal was achieved already by the absorbed dose of 30 kGy in the alkaline solution with the initial pH = 9.2 adjusted by Na_2CO_3 (at the dose of 70 kGy, pH of solution decreased to the value 7.6) (Figure 5). pH was adjusted also by NaOH to the value of 9 prior to irradiation and the system behaved uniformly as in the presence of CO_3^{2-} (including the decrease of pH). Therefore, CO_3^{2-} seems not to be the necessary component for the successful removal of lead independently of its ability to react with $^{\bullet}OH$ radicals.

Very interesting finding was that the same degree of lead removal can be achieved without addition of $^{\bullet}OH$ radical scavengers in the system containing Pb complexed with EDTA with initial natural pH = 3.2 (Figure 5) where the product is metallic lead.

In contrast to the behaviour of the alkaline system, pH was increasing and equal to 6 already by the absorption of the dose of 10 kGy (at this dose 80% degradation of EDTA already ocured as was proved by isotachoforesis and lead precipitated after the breaking of the complex (Figure 5)) and with further increasing dose the value shifted to 6.5. Moreover, the present ions NO_3^- behave as the scavenger of e^-_{aq} (Table 4) and hence contribute to the creation of the stronger oxidative environment suitable for the degradation of complexant. The thermogravimetric analysis showed that in the system with slight excess of EDTA metallic Pb with little amount of $PbCO_3$ was formed.

In a system with double concentration of EDTA compared to lead, the efficient removal of lead was achieved at a dose of 20 kGy higher. In a system with 1×10^{-2} mol.L^{-1} EDTA lead was not removed, even if the system was purged with N_2O (studied in the dose range 0-50 kGy). A significant negative impact of bubbling N_2 or N_2O (Figure 5) was found. If the system is bubbled with N_2O, already too oxidizing environment is formed that does not allow more effective reduction of lead.

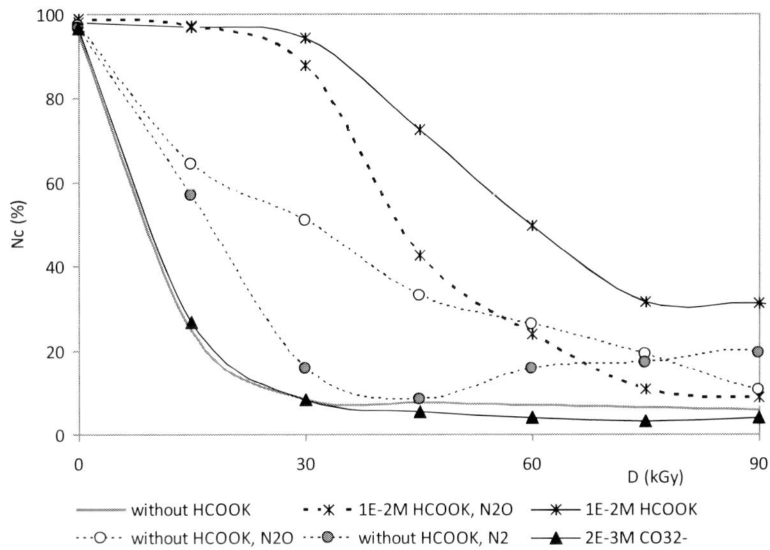

Figure 5. Dependence of normalized concentration N_C of Pb^{II} species on the dose of irradiation D (100 mg.L^{-1} of lead, 6×10^{-4} mol.L^{-1} EDTA) with or without an $^{\bullet}OH$ radical scavenger in systems bubbled with N_20 or N_2.

The $^{\bullet}OH$ radical scavengers HCOOK, 10% methanol and HCO_3^- affect radiation removal of lead complexed with EDTA interestingly. HCOOK in low concentration (1×10^{-3} mol.L^{-1}) does not have effect on the behavior of the system. On the other hand, the addition of HCOOK in higher concentration (1×10^{-2} mol.L^{-1}) has a strong negative effect as shown in Figure 5. Probably in this concentration, HCOOK already competes with complexant for the $^{\bullet}OH$ radicals. In the system with 10 % methanol no radiation removal of lead was observed up to 150 kGy. In the slightly alkaline system with 2×10^{-3} mol.L^{-1} HCO_3^-, a weaker $^{\bullet}OH$ radical scavenger (Table 4), the effect on the process is the same as in the system containing CO_3^{2-}.

The behaviour of system with lead dissolved from another salt (chloride or acetate) is different (Figure 6). It is very probably caused as in the case of added HCOOK by competition for $^{\bullet}OH$ radicals between complexant and anions of starting salts (Table 4).

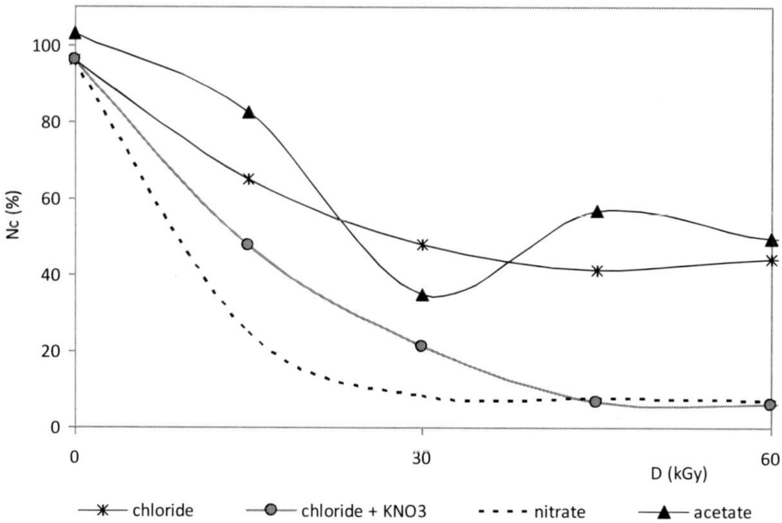

Figure 6. Dependence of normalized concentration N_C of Pb^{II} species on the dose of irradiation D (100 mg.L^{-1} of lead, 6×10^{-4} mol.L^{-1} EDTA, without $^{\bullet}OH$ radical scavenger) in systems containing different initial salts.

It is also supported by the behavior of system with Cl^- together with equimolar amount of NO_3^- (KNO_3) (Figure 6). It is evident that the presence of nitrate is significant for the process of radiation induced degradation.

The positive effect of the addition of solid phase has been observed mainly for Cu_2O in the standard solution ($Pb(NO_3)_2$) with addition of 1×10^{-2} mol.L^{-1} HCOOK. Cu_2O significantly improved removal efficiency of lead from the system, up to 40% (irrespective of the amount in the range from 0.25 to 1 mg.mL^{-1}). The solid is also added to the irradiated system with the same purpose as in the case of photocatalysis. The test was performed with ZrO_2, with the width of the band gap larger than that of TiO_2 (it should therefore respond better to irradiation). The system with 0.25 mg.mL^{-1} ZrO_2 behaved very similarly to the system of Cu_2O and the presumption was not proved. The addition of TiO_2 or Fe_2O_3 had positive effect but half in comparison with Cu_2O.

In system without scavenger of $^{\bullet}OH$ radicals, the additions of solids Fe_2O_3, ZrO_2, W_2O_3, TiO_2 including Cu_2O (Figure 7) caused no or a negative effect. Bubbling of these systems with N_2O deepened the negative effect.

Concerning cadmium complexed with EDTA (Drtinova 2013) was the best result 90 % of radiation induced precipitation of $CdCO_3$ (proved by X-ray diffraction) at a dose of 45 kGy with addition of 5×10^{-3} mol.L^{-1} carbonate (pH 10.5).

The presence of CO_3^{2-} seems to be necessary for the formation of the final precipitate because in the same system with pH adjusted by NaOH the radiation removal of Cd^{II} did not occur. Similarly to behavior of the system with lead, 1×10^{-2} or 1×10^{-3} mol.L^{-1} HCOOK inhibited the radiation removal of Cd^{II} (studied up to the absorbed dose of 100 kGy). In contrast to the behavior of system with lead, the removal of EDTA complexed cadmium did not occur with or without $^{\bullet}$OH radicals scavenger HCOOK and not even in a case of nitrate present as a cadmium salt anion (Table 4). In alkaline environment adjusted by adding bicarbonate (2×10^{-3} mol.L^{-1}) the radiation removal of cadmium was not observed.

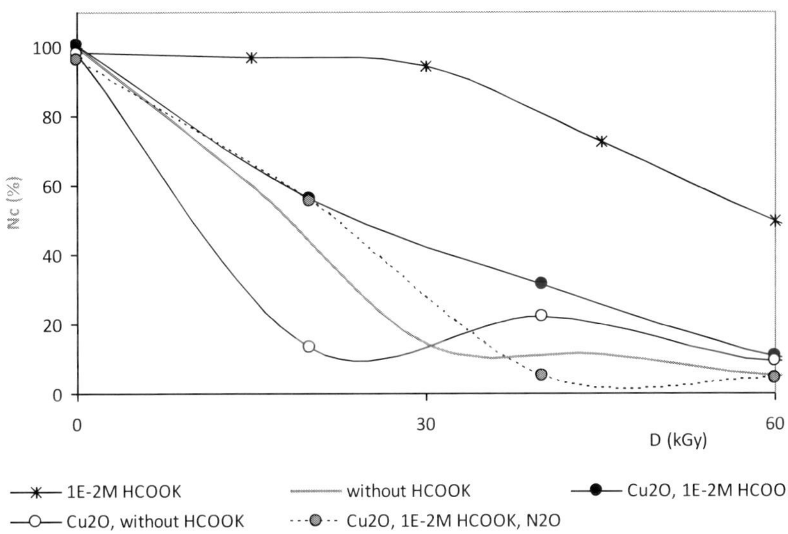

Figure 7. Dependence of normalized concentration N_C of Pb^{II} species on the dose of irradiation D (100 mg.L^{-1} of lead, 6×10^{-4} mol.L^{-1} EDTA) in the system containing HCOOK (1×10^{-2} mol.L^{-1}) and 0.25 mg.mL^{-1} Cu_2O.

Conclusion

The study of the influence of pH on the photocatalytic degradation of EDTA showed that the photosensitivity of EDTA depends on the degree of the dissociation of its molecule. Further, it proved the dependence of photocatalytic activity of TiO_2 catalysts on their isoelectric points. The „negative" photocatalytic activity observed at high pH values was ascribed to shielding of the EDTA solution by the particles of (at these pH) practically inactive photocatalysts.

In the tests focused on the comparison of various types of photocatalysts it was proved that not only the photocatalytic activity of the catalysts towards EDTA degradation varies with pH, but also that the behavior of the single photocatalyst differs significantly. Higher photocatalytic activities were found only for the samples TiO_2-M and Degussa P25, no simple correlation of their photocatalytic activity with their specific surface areas was found. It seemed that the method of the TiO_2 preparation influences the efficiency in more important way.

The experiments with metal complexes (Co, Cr) of EDTA showed that while pure EDTA can be degraded with high efficiency, the decomposition of complexes is much more complicated. The EDTA - Co complexes seem to be rather photoresistant, regardless of the type of the complex (CoH_2EDTA^0, $CoHEDTA^-$ or $CoEDTA^{2-}$). Any significant photodegradation proceeds on the surface of the photocatalyst, only. The EDTA - Cr complexes seem to be even more photoresistant, they cannot degrade on the surface of the photocatalyst, since no sorbable species exist in any pH range.

Degradation of EDTA via irradiation with accelerated electrons or γ-rays proved to be possible. Interestingly, even some heavy metals complexed with EDTA can be removed from aqueous solutions by this method where EDTA, at least partially, performs as an $^{\bullet}OH$ radicals scavenger. In the system with lead, the efficient degradation of complex with EDTA is achieved at a dose of 40 kGy up to 97 % without addition of any $^{\bullet}OH$ scavenger. The presence of nitrate (e^-_{aq} scavenger) is in this case important for an effective gradual reduction of complexed lead. Nearly the same efficiency in such wide range of pH (3.2 – 9.5) as stated above is reached in presence of $1x10^{-3}$ mol.L^{-1} HCOOK (pH about 5) or in an alkaline environment with $2x10^{-3}$ mol.L^{-1} CO_3^{2-} or $2x10^{-3}$ mol.L^{-1} HCO_3^-.

No positive effect from bubbling with nitrogen or oxygen was found. On the contrary, saturation with nitrous oxide in presence of scavengers has a positive effect by creating a stronger oxidizing environment and thus reducing the competition for $^{\bullet}OH$ radicals between complexant and scavenger. In the system containing cadmium, the mere presence of EDTA is not sufficient for the successful radiation removal of Cd and HCOOK as an $^{\bullet}OH$ radicals scavenger has to be added.

REFERENCES

Advances in Technologies for the Treatment of Low and Intermediate Level Radioactive Liquid Wastes; In Technical Report Series No.370; *IAEA*: Vienna, AU, 1994; 103 p.

Bednar, J. Theoretical Foundations of Radiation Chemistry; Academia Praha: Praha, CZ; 1990.

Bolton, J. R.; Valladares, J. E.; Cooper, W. J.; Waite, T. D.; Kurucz, C. N.; Nickelsen M. G.; Kajdi D. C. J. Adv Oxid Technol 1998, 3, 174 – 181.

Buitenhuis, R.; Bakker, C. M. N.; Stock, F. R.; Louwrier, P. W. F. *Radiochim. Acta.* 1977, 24, 189-192.

Buxton, G. V. In *Radiation chemistry: Principles and applications*; Farhataziz and Rodgers M. A. J.; Ed.; VCH Publishers: New York, US, 1987; pp 321-349.

Cernochova, K. *Separation of Radionuclides from Spent Decontamination Solutions*; Ph.D. Thesis, CTU: Prague, CZ, 2002; 148 p.

Chaychian, M.; Al-Sheikhly, M.; Silverman, J.; McLaughlin, W. L. *Radiat Phys Chem* 1998, 53, 145-150.

Clearfield, A. *Inorganic Ion Exchange materials*; CRC Press, Inc.: Boca Raton, FL, US, 1982.

Cubova, K.; Drtinova B. In: *Citric Acid: Synthesis, Properties and Applications;* Vargas, D.A. and Medina, J.V.; Ed.; Nova Science Publisher: Hauppauge, NY, US, 2012; pp 1-42.

Danacikova E. Degradation of free and Bound Complexants Using Oxidation and/or Photo-Oxidation Processes; Literature Review, CTU: Prague, CZ, 1995, 45 p.

Danacikova, E.; John, J.; Rosikova, K.; Sebesta, F.; Hooper, E.W. In *Proc. Workshop 98;* CTU: Prague, CZ, 1998; pp 819-820.

Danacikova, E.; John, J.; Sebesta, F.; Hooper, E.W. In *Proc. Workshop* 97;CTU: Prague, CZ, 1997; pp 155-156.

Danacikova E.; John J.; Sebesta F.; Motl A.; Hooper E.W. Czech. *J. Phys.* 1999, 49, S1 789-795.

Danacikova–Popelova, E.; John, J.; Rosikova, K.; Sebesta, F.; Hooper, E.W. In *Proc. Warsaw'98 Fourth International Symposium and Exhibition on Environmental Contamination in Central and Eastern Europe*; FSU: Tallahassee, FL, US, 1999.

Dolin, P. I.; Shubin, V. N.; Brusenceva, S. A. *Radiation Water Treatment*; Nauka: Moskva, SSSR, 1973.

Drtinova B. In: *Cadmuim*; Vargas, D.A. and Medina, J.V.; Ed.; Nova Science Publisher: Hauppauge, NY, US, 2013; (forthcoming).

Drtinova, B.; Pospisil, M. *J Radioanal Nucl* Ch 2010, 286(3), 631-636.

Drtinova*, B.; Pospisil, M.; Cuba, V. *Appl. Radiat. Isotopes*. 2010, 68(4-5), 672-675.

Drtinova, B.; Pospisil, M.; Cuba, V. In NRC7 CD Proceedings [CD-ROM], Hungarian Academy of Sciences, Budapest, HU, 2008, 5 pp.

Drtinova, B.; Pospisil, M.; Mucka, V.; Cuba, V. Czech *J. Phys*. 2006, 56, D381-387.

Getoff, N. *Res Chem Intermed*. 2001, 27(4, 5), 343-358.

Höbel, B.; Von Sonntag, C. *J. Chem. Soc*. 1998, Perkin Trans. 2, 509-513.

Hoffmann, M.R.; Martin, S.T.; Choi, W.; Bahnemann, D.W. *Chem. Rev*. 1995, 95, 69-96.

John, J.; Danacikova, E.; Sebesta, F., Motl, A.; Rosikova, K.; Hooper, *E.W. Chem. listy* 1997, 91(9), 770 – 771.

Kaiser, K.L.E. *Water Research* 1973, 7, 1465-1473.

Krapfenbauer, K.; Getoff, N. *Radiat. Physic. Chem*. 1999, 55, 385-393.

Legrini, O.; Oliveros, E.; Braun, A.M. *Chem. Rev*. 1993, 93, 671-698.

Litter, M.I.; Navio, J.A. J. *Photochem. Photobiol, A. Chem*. 1994, 84, 183-193.

Low, G. et al. *Environ. Sci. Technol*. 1991, 25, 460-467.

Malinka, E. A.; Kamalov, G. L. J. Photochem. Photobiolog. *A. Chem*. 1994, 81, 193-197.

Pope, M. The TiO_2 Photocatalysed Decomposition of Some Common Pollutants in Water and the Influence of Metal Ions on the Photocatalytic Process; Ph.D.Thesis; University of Reading: UK, 1998.

Popelova E. A Study of Radionuclides Extraction from Aqueous Systems after Degradation of Organic Complexants; Ph.D. Thesis, CTU: Prague, CZ, 2002, 129 p.

Pospisil, M.; Cuba, V.; Mucka, V.; Drtinova, B. *Radiat Phys Chem* 2006, 75(3), 403-407.

Rosikova K. Separation of Radionuclides from the Solutions Containing Organic Complexing Agents; Diploma Thesis, CTU: Prague, CZ, 1998, 75 p.

Rosikova, K., John, J., Danacikova–Popelova, E., Sebesta, F., Hooper, E.W. In *Proc. Warsaw'98 Fourth International Symposium and Exhibition on Environmental Contamination in Central and Eastern Europe*; FSU: Tallahassee, FL, US, 1999.

Rosikova, K.; John, J.; Sebesta F. *J. Radioanal. Nucl. Chem*. 2003, 255, 397.

Sabin, F.; Turk, T.; Vogler, *A. J. Photochem. Photobiol. A. Chem.* 1992, 63, 99-106.

Schiavello, M., Ed.; *Heterogeneous Photocatalysis; Wiley Series in Photoscience and Photoengineering*, Vol.3, J.Wiley and Sons: Chichester, UK, 1997; 197 pp.

Su, Y.; Wang, Y.; Daschbach, J. L.; Fryberger, T. B.; Henderson, M. A.; Janata, J., Peden, C. H. F. J. *Adv. Oxid. Technol.* 1998, 3(1), 63-69.

Swallow, A. J. Application of Pulse Radiolysis to the Study of Aqueous Organic Systems. *The Study of Fast Processes and Transient Species by Electron Pulse Radiolysis*; D. Reidel Publishing Company: Dordrecht, NL, 1982; pp 289 – 315.

Tanaka, K.; Hisanaga, T.; Harada, K. J. *Photochem. Photobiol., A. Chem.* 1989, 48, 155-159.

Visual Minteq, ver. 2.11. compiled by Gustafsson, KTH, Stockholm, Sweden; based on Minteq A2 ver. 4.0; Allisson et al., CEAM, EPA, USA, 1999.

Zhestkova, T. P.; Pikaev, A. A.; Pikaev, A. K.; Kim, D. K.; Kim, Y.; Han, B. *High Energ. Chem.*+ 1997, 31(4), 222-225.

In: EDTA
Editor: Andris Molnar

ISBN: 978-1-62808-146-6

Chapter 4

THE USE OF EDTA ROOT CONDITIONING IN REGENERATIVE PROCEDURES OF PERIODONTAL TISSUES

Daniela Leal Zandim-Barcelos [*]
and José Eduardo Cezar Sampaio [*]
Department of Diagnosis and Surgery, School of Dentistry at Araraquara, São Paulo State University – UNESP, Brazil

ABSTRACT

Periodontitis is a chronic destructive inflammatory disease caused by bacteria present in dental plaque. This condition induces the breakdown of tooth supporting apparatus, and the ultimate result of the untreated disease is tooth loss. The primary goal of periodontal therapy consists not only to arrest the tissue destruction caused by periodontal disease but also to regenerate the tooth supporting tissues, including alveolar bone, periodontal ligament and cementum. Several procedures have been proposed to enhance regeneration of periodontal tissues, including chemical root surface conditioning. The concept of root conditioning has been based on the perceived need for decontamination of the root surface devoid of attachment apparatus and for exposure of collagen fibers. Periodontitis-affected root surfaces are hypermineralized and contaminated with bacteria and inflammation-inducing bacterial components, such as endotoxin. In addition, the smear layer formed during mechanical root instrumentation has been suggested to act as a physical barrier, inhibiting new connective tissue attachment to the root surface. Therefore, a number of varying pH agents have been used to eliminate the factors that render the root surface bioincompatible for periodontal tissues regeneration. Root conditioning with a neutral pH agent such as EDTA appeared to produce in a short-term perspective a more biocompatible surface compared to conditioning at low pH. Etching at low pH may induce an immediate necrosis while etching at neutral pH preserves the integrity of exposed collagen fibers and the vitality of the adjacent tissues. EDTA root conditioning removes the root surface-associated smear layer, widening dentinal tubule orifices, and exposing collagen matrix when it is applied after scaling and root planing. The aim of this article is to present an overview of the biological effects of EDTA conditioning for root surface modification and to discuss the use of this chemical agent in regenerative procedures of periodontal tissues.

Introduction

Periodontitis is a chronic destructive inflammatory disease that occurs in response to bacterial accumulations on the teeth. The inflammation of periodontal tissues results in apical migration of junctional epithelium along the root surface and induces the breakdown of tooth supporting apparatus, including alveolar bone, periodontal ligament and cementum. The ultimate outcome of the untreated disease is tooth loss. Therefore, the periodontal therapy is primarily directed to halt the progression of periodontal attachment loss by removing etiologic factors in order to stabilize the long-term prognosis of the periodontium. Although it has been proven to be effective, the anti-infective therapy does not restore the supporting structures that were lost because of disease. This therapy results in periodontal repair in a way that the original architecture and function of the lost structures cannot be reestablished [1-3]. Regeneration requires restoration of alveolar bone height, regeneration of gingival connective tissue destroyed by inflammation, formation of new acellular extrinsic fiber cementum on previously exposed root surfaces, synthesis of Sharpey's fibers and their insertion into root surfaces, and re-establishment of epithelial seal at the coronal portion. However, the wound healing following conventional periodontal therapy has been characterized by development of long junctional epithelium and connective tissue adhesion [4, 5]. For this reason, predictable regeneration of periodontal tissues is considered a major challenge of periodontal therapy. Several procedures have been proposed to enhance regeneration of periodontal tissues, including chemical root surface conditioning, guided tissue regeneration, bone grafting, growth factors application or combination of these procedures [6].

During the progression of periodontal disease, root surface becomes exposed to the subgingival and/or oral environment as attachment loss occurs and progresses. The nature of periodontitis-affected root surface has been identified as one factor that negatively interferes with predictable periodontal regeneration [3, 7-9]. Pathologically exposed root surfaces have undergone substantial alterations and changes, such as denudation of fiber attachment, contamination by bacteria and their endotoxins, and alterations in mineral density and composition [10-16]. These alterations may have a significant biological influence on the potential of periodontal regeneration. The conventional treatment of periodontitis-affected root surfaces is based on the mechanical removal of plaque and calculus. However, the smear layer formed during mechanical root instrumentation has been suggested to act as a physical barrier, inhibiting new connective tissue attachment to the root surface and acting as a substrate for bacterial growth [17-19].

Since the root surface serves as a wound margin during regeneration, chemical conditioning has been directed to create a more biocompatible root surface for early cell migration, attachment, cell-matrix interaction, fiber development and, consequently, for periodontal regeneration. Root surface conditioning has been used to detoxify, to decontaminate, and to demineralize the root surface. This procedure modifies the root surface by removing smear layer and endotoxin, opening dentinal tubules and exposing the collagenous matrix [3, 17, 20-22]. Citric acid, tetracycline and ethylenediaminetetraacetic acid (EDTA) are the most widely used substances for this goal.

Of these chemical agents in clinical use, EDTA is the only one that exclusively exerts its demineralizing effect through chelating divalent cations at neutral pH. The use of a decalcifying agent that operates at a neutral pH, such as EDTA, not only preserves the vitality of the remaining periodontal cells close to the root surface, but also removes calcium ions from the collagenous dentin matrix more selectively than low-pH etching agents. Etching at neutral pH has been shown to preserve tissue vitality, while etching at low pH induces an immediate necrosis [23]. In addition, conditioning with low pH agents, such as citric acid and phosphoric acid, can dissolve or alter the structure of dentin associated collagen fibers, resulting in a granulated, rather than fibrous, dentin surface [24]. In fact, etching at neutral pH appears in the short-term perspective to produce a more biocompatible surface compared to etching at low pH [23,25,26].

The aim of this article is to provide an overview of the biological effects of EDTA conditioning for root surface modification and to discuss the use of this chemical agent in regenerative procedures of periodontal tissues.

SMEAR LAYER REMOVAL AND COLLAGEN EXPOSURE BY EDTA ROOT SURFACE CONDITIONING

In regenerative procedures, root surface conditioning has been recommended to remove smear layer and to expose collagenous matrix from root planed surfaces. Different chemical agents have been used to modify the root surface. The texture of dentin surfaces was determined after different etching procedures. Dentin surfaces were conditioned with citric and phosphoric acids as well as 24% EDTA [25]. It was verified that 24% EDTA operating at neutral pH was able to remove selectively mineral from dentin surface, exposing a collagenous matrix. This was in contrast to etching with the low pH agents. Citric and phosphoric acids removed not only the mineral component but also the collagenous matrix. EDTA-treated dentin surfaces appeared to be more inviting for cellular colonization and subsequent connective tissue formation than surfaces etched at low pH.

Blomlöf obtained similar results in another experiment [26]. EDTA demonstrated a higher capacity to expose selectively collagen fibers in both healthy cementum surfaces and periodontitis-affected dentin surfaces than agents operating at low pH, which seemed to erode the surfaces to varying degrees.

An in situ study was conducted to confirm the effect of EDTA conditioning on root surface morphology [27]. Periodontally involved human teeth scheduled for extraction were scaled and root planed or were scaled, root planed, and then etched with a 24% EDTA gel preparation before immediate extraction. The gel was applied with a syringe and, after 2 minutes, the roots were rinsed with saline and immediately extracted. After extraction, the teeth were immersed in a solution of 1% trypsin for 20 minutes. The root planed surfaces were covered by a homogenous smooth smear layer and did not present any patent dentinal tubule or collagen fibers exposure. In contrast, EDTA treated root surfaces revealed intertubular surfaces covered by a dense fibrillar network that extends into patent dentinal tubules.

Blomlöf and colleagues also evaluated the possibility of obtaining an acceptable smear layer removal and collagen exposure following EDTA conditioning with concentrations lower than the supersaturation (24%) [28]. Human dentin specimens were etched with the following concentrations of EDTA for 2 minutes: 1.5%, 5%, 15% and 24%. Neither 1.5% nor 5% EDTA etching was sufficient to dissolve the smear layer to any higher degree. Etching with 15% EDTA dissolved the smear more effectively than the lower concentrations and collagen fibers were sparsely found. The supersaturated 24% solution was significantly more effective than the lower concentrations of EDTA tested with regard to smear-removing capacity. Collagen fibers were seen covering the entire intertubular dentin surfaces following 24% EDTA etching. Based on these findings, EDTA should be used with the concentration between 15 to 24% in order to obtain an acceptable smear removing and collagen-exposing effect. However, the rate of dentin demineralization with a less than supersaturated solution of EDTA (<24%) can decrease significantly with time due to consumption of EDTA. Thus, a supersaturated EDTA solution may be necessary to achieve both smear removal and especially exposure of root-associated collagen fibers within a clinically acceptable time period [28].

The conditioning efficacy of different trademarks of 24% EDTA gel was evaluated by our research group, considering the variables time and application method, using scanning electron microscopy [29]. Human dentin specimens were conditioned with saline solution (control) or one of the following EDTA gel: 24% EDTA (Santa Paula Pharmacy), 24% EDTA-Texapon (Santa Paula Pharmacy); 24% EDTA (PrefGel – Straumann); 24% EDTA (Biodinamica). The gels were applied for 1, 2 or 3 minutes by topical application, vigorous or smooth friction. The 24% EDTA Biodinamica gel promoted less smear layer removal than the other groups (Figure 1). No significant differences could be detected among the different times of application. The smooth friction resulted in lower amount of residual smear layer. All 24% EDTA gels tested were able to remove smear layer and to expose dentinal tubules irrespective of application time; however, the application method by smooth friction was more effective than the other ones. The comparative analysis demonstrated that the Biodinamica gel was the least effective product, promoting less smear layer removal and exposure of collagen fibers than the other EDTA gels (Figure 1).

In general, studies demonstrate that 24% EDTA root conditioning is effective in removing the root surface smear layer, exposing and widening dentinal tubules, and promoting dentin-associated collagen matrix exposure when it is applied after scaling and root planing. EDTA conditioning has been found to be as effective as low pH agents without any morphological change on root surface. The gel preparation has been the most used vehicle of EDTA since it provides a better control of the etching agent.

Blood Clot Adhesion to EDTA Conditioned Root Surface

Periodontal regeneration is contingent on the adsorption and uninterrupted adhesion of a fibrin clot to a periodontally involved root surface. Clot adhesion appears vitally dependent on the formation of a resilient union between the clot and the root surface. Poor fibrin clot adhesion to root surfaces covered by instrumentation smear layer has been observed in vitro [30]. Root surface demineralization removes the smear layer exposing dentinal tubules and dentin collagen matrix, which supports adhesion of the fibrin clot during early wound healing events [31-33]. Fibrin linkage to exposed collagen fibrils on etched root surfaces prevented apical migration of epithelium and preceded collagen fiber attachment to the root surfaces [7].

Baker et al. (2005) evaluated clot adhesion to conditioned dentin surfaces [34]. Human dentin blocks were exposed to saturated citric acid solution or to 24% EDTA gel using standardized protocols. Fresh human whole blood was applied to the blocks. After scanning electron microscopy analysis, it was verified that citric acid removed the dentin smear layer, exposing dentinal tubules and collagen fibers. EDTA appeared to be less efficacious than citric acid, leaving smear layer residues. Citric acid treated dentin surface provided the best support for fibrin clot adhesion. Forces produced by the rinse protocol partially removed the fibrin clot from EDTA-treated surfaces. In fact, citric acid surface demineralization removed the dentin smear layer to promote adhesion of a fibrin clot. On the other hand, EDTA gel was less effective than citric acid to promote adhesion of a fibrin clot. This effect could negatively influence the progression of wound maturation into a connective tissue attachment in vivo.

Using a similar protocol, our research group [35] compared the adhesion and maturation of blood components on chemically conditioned root surfaces. Human root specimens were manually scaled and one of the following treatments was applied: saline solution (control); 24% EDTA gel; 25% citric acid solution; tetracycline solution (50 mg/mL) or 30% sodium citrate solution. After treatment, each specimen received a blood drop. Photomicrographs were analyzed according to the score of adhesion of blood components. Control group was characterized by the absence of blood elements on root surface (Figure 2A).

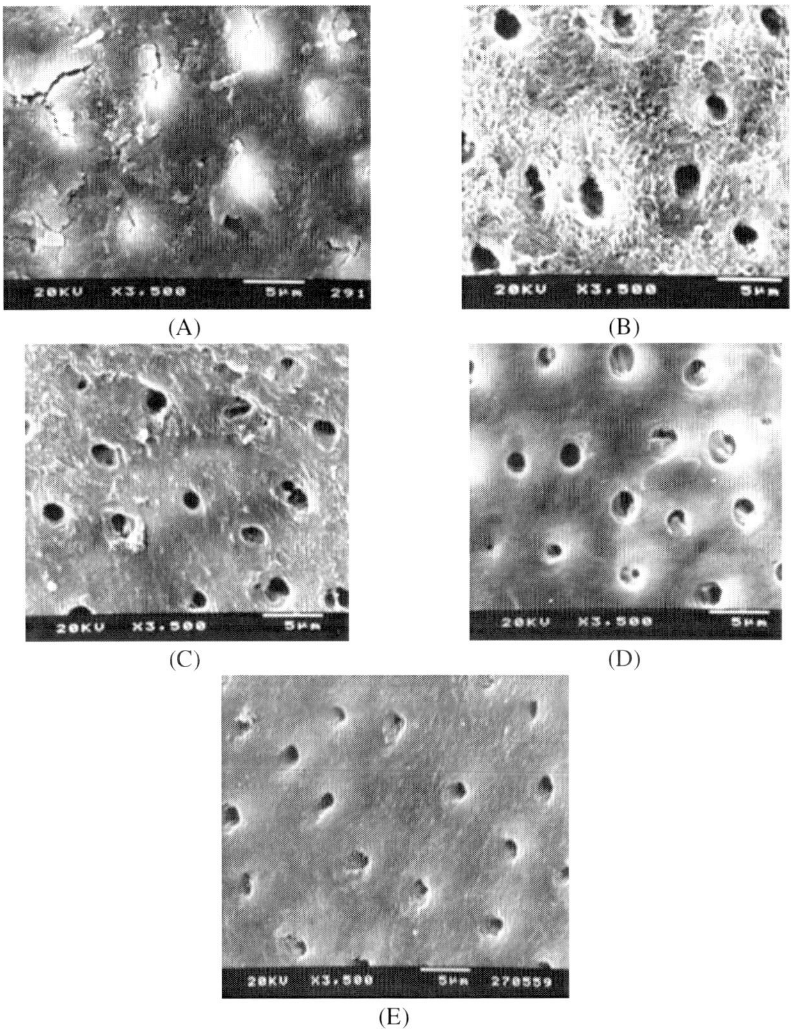

Figure 1. Representative photomicrographs of root surface morphology after 24% EDTA conditioning. (A) Saline solution irrigation (control). (B) 24% EDTA (PrefGel – Straumann). (C) 24% EDTA (Santa Paula Pharmacy). (D) 24% EDTA-Texapon (Santa Paula Pharmacy). (E) 24% EDTA (Biodinamica). The 24% EDTA gels were applied by smooth friction for 2 minutes. (bar = 5 μm; original magnification x3,500). Courtesy of Dr. José Eduardo Cezar Sampaio.

The best result was observed in citric acid group where a dense fibrin network with blood elements was adhered (Figure 2C). In EDTA group, a moderate fibrin network formation was observed (Figure 2B). In contrast, scarce fibrin network and few cells were detected in tetracycline samples (Figure 2D) and absence of blood elements was found on sodium citrate specimens (Figure 2E). Citric acid group was statistically different from control group. No differences were found among control, EDTA, tetracycline and sodium citrate groups. Under these experimental conditions, citric acid can be considered the indicated demineralizing agent to stabilize clot on the root surface that acts as a scaffold for connective tissue cells development. EDTA could have inhibited blood elements adsorption and adhesion to dentin surface because of a possible incomplete removal from root surface. EDTA is a calcium chelator that operates at neutral pH; thus, its residues could have inhibited or retarded coagulation events.

Cells and Biologically Active Substances Adhesion to EDTA Conditioned Root Surface

An appropriate root surface for cell attachment and orientation is essential for periodontal regeneration. Experimental studies indicate that fibroblasts adhesion to the root surfaces seems to be a critical step in the formation of a new connective tissue attachment [36, 37]. Aleo et al. [38] demonstrated that human gingival fibroblasts attachment did not occur to periodontitis-affected root surfaces, whereas attachment occurred to non-periodontitis or normal root surfaces. Therefore, root conditioning may be a necessary prerequisite for creating a suitable substrate for cell attachment and migration.

The study of Pitaru & Melcher [37] evaluated the influence of demineralization on the migration, attachment, and orientation of human gingival fibroblasts to human dental root surfaces in vitro. Porcine root slices were placed onto tissue culture dishes with a distance of 0.5 mm. Half of the root slices was demineralized with 18% EDTA, and the other group served as untreated controls. Cell migration and attachment were examined for periods up to 30 days. Cells migrated towards and attached to the root slices almost immediately following culturing of the fibroblasts with the specimens. Demineralized surfaces enabled the attachment, migration and contraction of fibroblasts, which led to the development of an oriented fiber attachment system on demineralized surfaces.

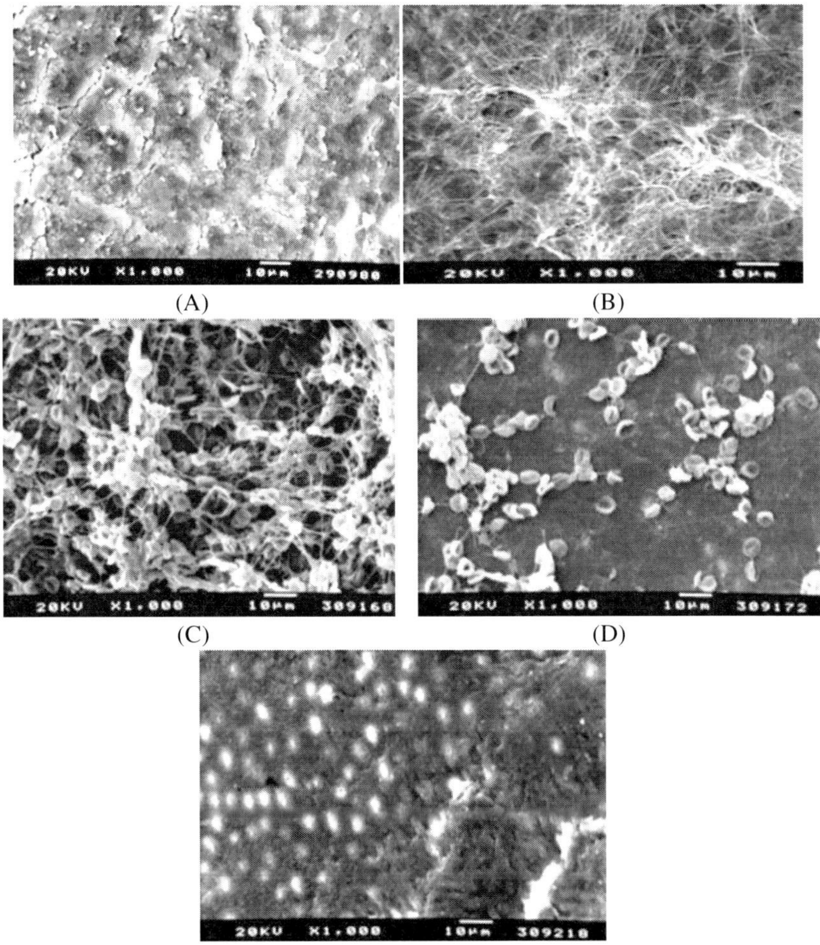

Figure 2. Adhesion and maturation of blood components after root surface conditioning. (A) Saline solution irrigation (control) - absence of blood elements on root surface. (B) 24% EDTA gel - root surface covered by a moderate layer of blood cells on the fibrin network. (C) 25% citric acid solution - root surface covered by a dense layer of blood cells on the fibrin network. (D) 50 mg/mL tetracycline solution - root surface covered by a scarce layer of blood cells on the fibrin network. (E) 30% sodium citrate solution - absence of blood elements on root surface (bar = 10 μm; original magnification x1,000). Courtesy of Dr. José Eduardo Cezar Sampaio.

The early responses of human periodontal ligament cells attached to EDTA-demineralized dentin surfaces treated with recombinant human platelet-derived growth factor-BB (rhPDGF-BB) or bone morphogenetic protein-2 (rhBMP-2) were investigated [39]. Dentin blocks were prepared from periodontally diseased human tooth roots. After demineralization with 24% EDTA, dentin blocks were treated with rhPDGF-BB, rhBMP-2 or only minimum essential medium (control). Human periodontal ligament cells were seeded on dentin surfaces and incubated. The alkaline phosphatase (ALP) activity and protein concentration of the attached cell were assessed at 2, 4 and 7 days. A significant increase of protein concentration and cell number was observed in PDGF-BB treated groups compared to control; however, the same increase was not verified for the ALP activity. In addition, a significant increase of ALP activity was detected in BMP-2 treated groups compared to control, but protein concentration and cell number remained almost the same over time. Thus, rhPDGF-BB and rhBMP-2 application to EDTA conditioned dentin surfaces promoted the early human periodontal ligament cell responses by increasing cell proliferation and differentiation, enhancing periodontal regeneration.

The same research group [40] evaluated the attachment of human periodontal ligament cells to periodontally diseased root surfaces after demineralization with low and neutral pH etching solutions. After scaling and root planing, root surfaces were conditioned with citric acid (pH 1.2), tetracycline hydrochloride (100 mg/ml), 24% EDTA or received no treatment. Samples were cultured with human periodontal ligament cells for 1, 3 and 7 days. A significantly higher cell attachment and orientation index to demineralized dentin surfaces with EDTA and citric acid was observed when compared to non-etched dentin surfaces. EDTA and citric acid demineralized human root surfaces provided more favorable surface for human periodontal ligament migration, attachment and orientation than the scaled and root planed diseased root surface alone.

Gamal and Mailhot (2003) investigated the correlation between root surface changes after EDTA conditioning and periodontal ligament fibroblast adhesion [41]. Periodontally-affected teeth were scaled and root planed or scaled, root planed and conditioned with 24% EDTA gel prior to immediate extraction. Half of the samples was cultured with periodontal ligament fibroblast for 24 hours before microscopic analysis. Root surfaces that were only instrumented revealed the presence of smear layer covering completely the orifices of the underlying dentinal tubules. EDTA gel exposure for 2 minutes following scaling and root planing resulted in smear layer removal and marked exposure of dentinal tubules openings. The root surfaces after 4-minute EDTA conditioning had a fibrillar texture associated with an increased diameter of the exposed dentinal tubules. With regard to fibroblasts adhesion, samples conditioned with EDTA for 2 minutes showed a significant increase in the number of flat and round adherent cells when compared to the non-etched samples, and a significant lower number of flat cells when compared to the 4-minute etched samples. In fact, 24% EDTA conditioning demonstrated a high capacity to enhance fibroblasts adhesion when it was applied after scaling and root planing. A positive correlation was found between time of EDTA gel conditioning and the degree of fibroblast adhesion.

Growth factors are considered biologic mediators due to their ability to regulate actively various functions of cells. Research on the application of tissue growth factors for promoting periodontal regeneration is still receiving increasing attention. It is postulated that the use of EDTA gel as a root surface pre-conditioning agent could potentiate and/or support the effects of using growth factor as a promising biologic root surface mediator in the healing process of periodontal tissues. A study was conducted to evaluate the effect of root conditioning with basic fibroblast growth factor (b-FGF) on morphology and proliferation of fibroblasts [42]. Dentin samples were previously etched with tetracycline-hydrochloride (50 mg/ml) or 24% EDTA and were treated with b-FGF (with 50 or 125 μg/ml). Non-treated dentin samples were used as control. Root surfaces treated with 125 μg/ml b-FGF and previously conditioned with tetracycline or EDTA presented morphology suggestive of cellular adhesion and viability. The density of fibroblasts on samples previously conditioned with EDTA, regardless of treatment with b-FGF, was significantly higher than in the other groups. The present findings suggest that topical application of b-FGF has a positive influence on both the density and morphology of fibroblasts.

Miyaji et al. (2010) investigated in vivo the cementum-like tissue deposition on bone morphogenetic protein (BMP) conditioned root surfaces pre-demineralized with 24% EDTA [43]. Demineralized dentin surfaces were treated with saline or BMP-2 solution (0.4 and 1.0 μg/μL). Formation of cementum-like tissue was frequently observed on the BMP-2 conditioned root at the coronal portion. Root surfaces treated with BMP-2 at 1.0 μg/μL showed significantly higher cementum-like tissue formation than control surfaces and root surfaces treated with BMP-2 at 0.4 μg/μL. Downgrowth of the junctional epithelium in the 1.0 μg/μL BMP-2 group was significantly less than that in the control roots. Root surface conditioning with BMP-2 after EDTA etching stimulated cementum-like tissue formation and inhibited epithelial downgrowth.

The study of Belal et al. (2010) verified the effect of EDTA and/or platelet-derived growth factor (PDGF) application on root adhesion and proliferation of periodontal ligament fibroblast cells [44]. Specimens prepared from periodontitis teeth received one of the following treatments: no treatment, scaling and root planing, 24% EDTA conditioning, PDGF (25 ng/ml) application or combined application of EDTA and PDGF. All groups, including untreated specimens, showed significantly increase of adhered cells from baseline to 7 days. However, the rate of increase was considerable higher in EDTA, PDGF, and combined groups. On day 7, PDGF and combined groups showed significantly higher number of adhesion cells than that found in the diseased, scaled and root planed and EDTA groups. Thus, root conditioning with EDTA enhanced cell adhesion more than scaling and root planing alone. There was no significant difference of cell number between PDGF and combined group. Combined application of EDTA and PDGF increased significantly periodontal ligament fibroblast cells adhesion than EDTA alone. PDGF alone, however, showed comparable effect to the combined application at all periods.

The effect of EDTA root surface conditioning on adhesion of clot blended β-tricalcium phosphate (β-TCP) bone substitute was also evaluated [45]. Sixty patients with chronic periodontitis and two teeth designated for extraction were included in the study. Intrabony defects were grafted with β-TCP or were previously etched with 24% EDTA gel followed by defect overfill with β-TCP. For each patient, one tooth was extracted immediately after surgery and the other tooth was extracted at 14 days for scanning electron microscopy evaluation. After 14 days, samples only grafted with β-TCP revealed undetected graft particles while samples pre-conditioned with EDTA showed areas of adherent β-TCP particles covering about 42.5 to 82.7% according to the particle size of β-TCP and EDTA exposure time used. The results demonstrated that root conditioning with EDTA gel improves β-TCP blended clot adhesion to periodontally involved root surfaces. The improvement of graft adhesion is supposed to retard the apical migration of epithelial attachment, a factor that could enhance periodontal regeneration during the early stages of healing.

ANIMAL STUDIES USING EDTA CONDITIONING OF ROOT SURFACE

Periodontal healing depends on viability and vitality of the tissues surrounding the damaged area of periodontium. For this reason, Blomlöf and Lindskog (1995) investigated the necrotizing effects on periodontal tissues of etching agents operating at low pH or neutral pH [25]. Phosphoric acid (37%) and citric acid (pH 1.0), both operating at low pH, exerted immediate necrotizing effects on mucosal flaps as well as on periodontal tissues. The penetration depth of phosphoric acid increased with time up to ¼ of the circumference of the root after 3 minutes. Citric acid treated specimens presented similar pattern with the exception of lower penetration depth of the acid effect, about half of the depth observed for phosphoric acid etched specimens. In contrast, 24% EDTA, which operates at neutral pH, did not induce any detectable necrosis during the experimental periods.

Periodontal healing of dehiscence defects after 24% EDTA conditioning (neutral pH agent) was compared to citric acid (pH 1.0) etching [27]. Defects surgically created in monkeys teeth were treated with or without root surface conditioning. Healing results were evaluated after 8 weeks. The statistically significant differences between EDTA treated surfaces and control surfaces were approximately 10% less failure (gingival recession and periodontal pocket), 10 to 15% more total histological attachment (long epithelial junction, connective tissue and reparative cementum), approximately 20% less long epithelial junction and 20% more connective tissue in roots etched with EDTA. On the other hand, the statistically significant differences between citric acid and control surfaces were approximately 10% more connective tissue and 15% less long epithelial junction in the citric acid etched surfaces. Etching at low pH improved healing compared with matched controls, although not to the same extent as etching at neutral pH. EDTA etching appeared to improve healing, avoiding the superficial necrotizing effect on periodontal tissues by citric acid and promoting exposure of more intact collagen fibers.

Our research group investigated the periodontal healing pattern of dehiscence-type defects following chemical root surface conditioning [46]. Defects were treated with sterile saline solution (control group) or one chemical conditioning modality: citric acid (pH 1.3), 24% EDTA, tetracycline hydrochloride (50 mg/ml), citric acid + tetracycline, or tetracycline + citric acid. After three months of healing, all treatments yielded significant improvements in terms of probing depth decrease and clinical attachment level gain compared to baseline values; however, without significant differences among the groups. Histometric analyses revealed that the highest amount of new cementum occurred in the EDTA etched group (3.72 ± 0.83 mm); while the lowest amount of new bone was observed in the group treated with only tetracycline (0.7 ± 0.94 mm). However, no statistically significant differences could be observed among the groups regarding epithelial apical migration, new cementum and alveolar bone formation. In fact, chemical root surface conditioning did not promote any significant improvement in periodontal healing pattern of dehiscence-type defects in dogs.

CLINICAL STUDIES USING EDTA CONDITIONING OF ROOT SURFACE

EDTA etching was found to be as effective as low pH etchants with regard to smear removal and superior in exposing root surface-associated collagen [23, 28]. However, studies failed to demonstrate clinical advantages of EDTA gel as an adjunctive therapy to surgical or non-surgical therapy.

Mayfield et al. (1998) evaluated the clinical treatment effect of 24% EDTA root conditioning in combination with surgical periodontal therapy in humans [47]. A comparison was made with conventional surgical flap therapy alone for the treatment of intraosseous periodontal defects. After 6 months of healing, probing attachment level gain for the EDTA group was 1.8 ± 1.5 mm and for the control group was 1.0 ± 1.7 mm. A probing bone gain of 1.0 ± 1.3 mm was found in the EDTA group and a non-significant gain of 0.4 ± 1.2 mm was measured in the control group. No statistically significant differences in clinical parameters could be detected between the two treatment groups with exception of gingival recession. EDTA group had a mean recession of 1.1 ± 0.8 mm while the control group had a mean recession of 1.8 ± 0.7 mm.

Blomlöf et al. (2000) investigated the effect of EDTA application as an adjunctive therapy to subgingival root debridement [48]. Sites with an initial probing pocket depth of 5 mm or more were treated with subgingival root planing followed by saline rinsing (control) or were treated with subgingival root planing in addition to EDTA etching of the root surface. The 24% EDTA gel preparation was applied with a syringe. After 6 months of treatment, the mean attachment gain and pocket reduction were respectively 0.76 ± 0.92 mm and 1.68 ± 1.38 mm for the control group, and 0.61 ± 1.04 mm and 1.37 ± 1.02

mm for the etched group. The results failed to demonstrate any significant difference between control and EDTA treatments with respect to probing pocket depth reduction and clinical attachment gain. EDTA application failed to add to the clinical healing result when combined with conventional nonsurgical periodontal therapy.

The same group evaluated the effect of 24% EDTA conditioning on surgical periodontal therapy. For comparison, citric acid (pH 1.0) treatment and control surgery only were included [22]. No statistically significant differences among treatment groups concerning attachment gain and pocket depth reduction could be seen after 3 and 6 months. Mean clinical attachment gain and probing depth reduction 6 months after treatment were respectively 1.30 ± 1.11 and 3.07 ± 1.01 mm for the control group, 1.32 ± 1.40 and 3.10 ± 1.56 mm for the group etched with citric acid, 0.99 ± 1.40 and 2.54 ± 1.34 mm for the EDTA etched group.

Mariotti performed a systematic review to assess the efficacy of root surface conditioning with citric acid, tetracycline, or EDTA in patients with chronic periodontitis [49]. Meta-analysis performed on 28 clinical trials showed no significant effect of root conditioning on attachment level gain and probing depth reduction. Thus, the evidence suggested that the use of chemical root surface modifiers provides no clinically significant benefit to regeneration in patients with chronic periodontitis. It was highlighted that following research should elucidate the biological effects of root surface modifiers on wound healing to determine future subclinical or incremental effects that may promote periodontal regeneration.

Root Surface Conditioning with EDTA Prior to Enamel Matrix Derivative Application

Enamel matrix derivative (EMD) was developed after the discovery of enamel matrix proteins and their involvement in cementum formation. EMD (Emdogain®) is derived from porcine teeth and mainly composed of enamel matrix proteins (amelogenins > 90%). This biological agent has been used as an adjunct to periodontal surgery to support periodontal tissues regeneration. Histologic evidence from animal and human studies revealed that topical application of EMD onto exposed root surfaces can induce new cementum, periodontal ligament, and alveolar bone formation in various types of periodontal defects [50-54]. The clinical protocol for the use of EMD includes root surface conditioning with 24% EDTA gel at neutral pH immediately prior to EMD application.

A recent systematic review examined the benefit of using EMD in the treatment of intrabony defects, furcation defects, and recession coverage [55]. After screening, 27 studies (20 for intrabony defects, one for furcation, and six for recession) were eligible for the review. A meta-analysis was performed for intrabony defects and recession. In the treatment of intrabony defects, the use of EMD showed significantly more clinical attachment gain (1.30 mm) and probing depth reduction (0.92 mm) than open-flap debridement, EDTA conditioning, or placebo treatment. However, EMD was equally effective as resorbable graft membrane in intrabony defects treatment considering clinical attachment gain and probing depth reduction. Only one randomized clinical trial compared the use of EMD to a membrane in furcation defects [56]. The EMD group showed significantly more reduction in horizontal furcation depth (2.6 - 1.8 mm) and less recession and postoperative complications than the group treated with resorbable membranes (1.9 - 1.4 mm). The use of EMD in combination with a coronally advanced flap showed significantly more complete root coverage compared with a coronally advanced flap alone (odds ratio of 3.5); however, the result was not significantly different when compared with a connective tissue graft.

Final Considerations

Chemical root surface conditioning has been employed in an effort to improve the clinical and histological regenerative results after periodontal treatment. Although the use of conditioning agents could be considered the oldest and most frequently attempted type of periodontal regeneration, controversies exist concerning the benefits of root surface conditioning as an adjunct to periodontal therapy. Root conditioning is considered an effective procedure to remove root surface smear layer and to promote collagen fibrils exposure within cementum or dentin matrix. EDTA conditioning has been found to be as effective as low pH agents in removing the smear layer and promoting collagen matrix exposure, without any morphological change on root surface. However, no additional improvement in clinical parameters has been indicated from clinical trials using any type of conditioning agent. The use of EDTA to modify the root surface provided no benefit of clinical significance to regeneration in patients with chronic periodontitis. Although no advantage could be observed with root surface conditioning alone, the exposure of collagenous matrix of dentin or cementum could be useful to retain biologically active substances such as growth factors, i.e. enamel matrix proteins, in regenerative procedures. Therefore, the potential contribution of EDTA root conditioning in combination with growth factor therapy remains to be determined.

References

[1] Albair, W. B., Cobb, C. M. & Killoy, W. J. (1982). Connective tissue attachment to periodontally diseased roots after citric acid demineralization. *Journal of Periodontology*, *53*, 515-526.

[2] Caton, J. G. & Zander, H. A. (1979). The attachment between tooth and gingival tissues after periodic root planing and soft tissue curettage. *Journal of Periodontology*, *50*, 462-466.

[3] Lowenguth, R. A. & Blieden, T. M. (1993). Periodontal regeneration: root surface demineralization. *Periodontology 2000*, *1*, 54-68.

[4] Caton, J. & Nyman, S. (1980). Histometric evaluation of periodontal surgery. I. The modified Widman flap procedure. *Journal of Clinical Periodontology*, *7*, 212-223.

[5] Caton, J., Nyman, S. & Zander, H. (1980). Histometric evaluation of periodontal surgery. II. Connective tissue attachment levels after four regenerative procedures. *Journal of Clinical Periodontology*, *7*, 224-231.

[6] Wikesjö, U. M. & Selvig, K. A. (1999). Periodontal wound healing and regeneration. *Periodontology 2000*, *19*, 21-39.

[7] Polson, A. M. & Caton, J. (1982). Factors influencing periodontal repair and regeneration. *Journal of Periodontology*, *53*, 617-625.

[8] Proye, M. P. & Polson, A. M. (1982). Effect of root surface alterations on periodontal healing. I. Surgical denudation. *Journal of Clinical Periodontology*, *9*, 428-440.

[9] López, N. J. (1984). Connective tissue regeneration to periodontally diseased roots, planed and conditioned with citric acid and implanted into the oral mucosa. *Journal of Periodontology*, *55*, 381-390.

[10] Hatfield, C. G. & Baumhammers, A. (1971). Cytotoxic effects of periodontally involved surfaces of human teeth. *Archives of Oral Biology*, *16*, 465-468.

[11] Aleo, J. J., De Renzis, F. A., Farber, P. A. & Varboncoeur, A. P. (1974). The presence and biologic activity of cementum-bound endotoxin. *Journal of Periodontology*, *45*, 672-675.

[12] Adriaens, P. A., De Boever, J. A. & Loesche, W. J. (1988). Bacterial invasion in root cementum and radicular dentin of periodontally diseased teeth in humans. A reservoir of periodontopathic bacteria. *Journal of Periodontology*, *59*, 222-230.

[13] Selvig, K. A. (1969). Biological changes at the tooth saliva interface in periodontal disease. *Journal of Dental Research*, *48*, 846-855.

[14] Selvig, K. A., Bogle, G. C., Sigurdsson, T. J. & Wikesjö, U. M. (1996). Does root surface conditioning with citric acid delay healing? *Journal of Clinical Periodontology*, *23*, 119-127.

[15] Selvig, K. A. & Hals, D. (1977). Periodontally diseased cementum studied by correlated microradiography, electron probe analysis and electron microscopy. *Journal of Periodontal Research*, *12*, 419-429.

[16] Socransky, S. S. & Haffajee, A. D. (1997). The nature of periodontal diseases. *Annals of Periodontology*, *2*, 3-10.

[17] Polson, A. M., Frederick, G. T., Ladenheim, S. & Hanes, P. J. (1984). The production of a root surface smear layer by instrumentation and its removal by citric acid. *Journal of Periodontology*, *55*, 443-446.

[18] Hanes, P., Polson, A. & Frederick, T. (1991). Citric acid treatment of periodontitis-affected cementum. A scanning electron microscopic study. *Journal of Clinical Periodontology*, *18*, 567-575.

[19] Nalbandian, J. & Cote, N. (1982). Direct histological comparison of periodontal wound healing in the beagle dog with and without citric acid conditioning. *Journal of Periodontal Research*, *17*, 552-562.

[20] Lasho, D. J., O'Leary, T. J. & Kafrawy, A. H. (1983). A scanning electron microscope study of the effects of various agents on instrumented periodontally involved root surfaces. *Journal of Periodontology*, *54*, 210-220.

[21] Klinge, B. (1997). Root surface conditioning. In: Lang NP, Karring T, Lindhe *J. Proceedings of the 2nd European Workshop on Periodontology*. Berlin: Quintessence, 276-283.

[22] Blomlöf, L., Jonsson, B., Blomlöf, J. & Lindskog, S. (2000). A clinical study of root surface conditioning with an EDTA gel. II. Surgical periodontal treatment. *The International Journal of Periodontics and Restorative Dentistry*, *20*, 566-573.

[23] Blomlöf, J. & Lindskog, S. (1995). Periodontal tissue – Vitality after different etching modalities. *Journal of Clinical Periodontology*, *22*, 464-468.

[24] Van Meerbeek, B., Dhem, A., Goret-Nicaise, M., Braem, M., Lambrechts, P. & Vanherle, G. (1993). Comparative SEM and TEM

examination of the ultrastructure of the resin-dentin interdiffusion zone. *Journal of Dental Research*, *72*, 495-501.

[25] Blomlöf, J. & Lindskog, S. (1995). Root surface texture and early cell and tissue colonization after different etching modalities. *European Journal of Oral Sciences*, *103*, 17-24.

[26] Blomlöf, J. (1996). Root cementum appearance in healthy monkeys and periodontitis-prone patients after different etching modalities. *Journal of Clinical Periodontology*, *23*, 12-18.

[27] Blomlöf, J., Jansson, L., Blomlöf, L. & Lindskog, S. (1996). Root surface etching at neutral pH promotes periodontal healing. *Journal of Clinical Periodontology*, *23*, 50-55.

[28] Blomlöf, J., Blomlöf, L. & Lindskog, S. (1997). Effect of different concentrations of EDTA on smear removal and collagen exposure in periodontitis-affected root surfaces. *Journal of Clinical Periodontology*, *24*, 534-537.

[29] Sousa, C. P., Frizzera, F., Batista, L. H. C., Dantas, A. A. D., Zandim-Barcelos, D. L. & Sampaio, J. E. C. (2013). Comparação in vitro da eficácia de diferentes formulações do gel de EDTA 24% no condicionamento da superfície radicular. *Revista de Odontologia da UNESP*, *42*, 7-12.

[30] Baker, P. J., Rotch, H. A., Trombelli, L. & Wikesjö, U. M. E. (2000). An in vitro screening model to evaluate root conditioning protocols for periodontal regenerative procedures. *Journal of Periodontology*, *71*, 1139-1143.

[31] Hanes, P. J., Polson, A. M. & Ladenheim, S. (1985). Cell and fiber attachment to demineralized dentin from normal root surfaces. *Journal of Periodontology*, *56*, 752-765.

[32] Polson, A. M. & Proye, M. P. (1983). Fibrin linkage: a precursor for new attachment. *Journal of Periodontology*, *54*, 141-147.

[33] Wikesjö, U. M., Nilvéus, R. E. & Selvig, K. A. (1992). Significance of early healing events on periodontal repair: a review. *Journal of Periodontology*, 63, 158-165.

[34] Baker, D. L., Stanley Pavlow, S. A. & Wikesjö, U. M. E. (2005). Fibrin clot adhesion to dentin conditioned with protein constructs: an in vitro proof-of-principle study. *Journal of Clinical Periodontology*, *32*, 561-566.

[35] Leite, F. R. M., Sampaio, J. E. C., Zandim, D. L., Dantas, A. A. R., Leite, E. R. M. & Leite, A. A. (2009). Root surface conditioning with

acid and chelating agents influence clot stabilization. *Quintessence International*, *41*, 341-349.

[36] Pitaru, S., Gray, A., Aubin, J. E. & Melcher, A. H. (1984). The influence of the morphological and chemical *Journal of Periodontal Research*, 19, 408-418.

[37] Pitaru, S. & Melcher, A. H. (1987). Organization of an oriented fiber *Journal of Periodontal Research*, *22*, 6-13.

[38] Aleo, J. J., DeRenzis, F. A. & Farber, P. A. (1975). In vitro attachment of human gingival fibroblasts to root surfaces. *Journal of Periodontology*, *46*, 639-345.

[39] Zaman, K. U., Sugaya, T. & Kato, H. (1999). Effect of recombinant human platelet-derived growth factor-BB and bone morphogenetic protein-2 application to demineralized dentin on early periodontal ligament cell response. *Journal of Periodontal Research*, *34*, 244-250.

[40] Zaman, K. U., Sugaya, T., Hongo, O. & Kato, H. (2000). A study of attached and oriented human periodontal ligament cells to periodontally diseased cementum and dentin after demineralizing with neutral and low pH etching solution. *Journal of Periodontology*, *71*, 1094-1099.

[41] Gamal, A. Y. & Mailhot, J. M. (2003). The effects of EDTA gel conditioning exposure time on periodontitis-affected human root surfaces: surface topography and PDL cell adhesion. *Journal of The International Academy of Periodontology*, *5*, 11-22.

[42] Silvério, K. G., Martinez, A. E. & Rossa, C. Jr. (2007). Effects of basic fibroblast growth factor on density and morphology of fibroblasts grown on root surfaces with or without conditioning with tetracycline or EDTA. *Journal of Oral Science*, *49*, 213-220.

[43] Miyaji, H., Sugaya, T., Ibe, K., Ishizuka, R., Tokunaga, K. & Kawanami, M. (2010). Root surface conditioning with bone morphogenetic protein-2 facilitates cementum-like tissue deposition in beagle dogs. *Journal of Periodontal Research*, *45*, 658-663.

[44] Belal, M. H., Watanabe, H., Ichinose, S. & Ishikawa, I. (2012). Effect of PDGF-BB combined with EDTA gel on adhesion and proliferation to the root surface. *Odontology / the Society of the Nippon Dental University*, 100, 206-214.

[45] Gamal, A. Y. (2011). Enhanced β-tricalcium phosphate blended clot adhesion to EDTA biomodulated periodontally affected root surfaces: in vivo scanning electron microscopy evaluation. *Journal of Periodontology*, *82*, 1587-1595.

[46] Zandim, D. L., Leite, F. R., da Silva, V. C., Lopes, B. M., Spolidorio, L.C. & Sampaio, J. E. (2012). Wound healing *Clinical Oral Investigations*. Epub ahead of print. DOI 10.1007/s00784-012-0848-4

[47] Mayfield, L., Söderholm, G., Norderyd, O. & Attström, R. (1998). Root conditioning using EDTA gel as an adjunct to surgical therapy for the treatment of intraosseous periodontal defects. *Journal of Clinical Periodontology*, *25*, 707-714.

[48] Blomlöf, L., Bergman, E., Forsgardh, A., Foss, L., Larsson, A., Sjöberg, B., Uhlander, L., Jonsson, B., Blomlöf, J. & Lindskog, S. (2000). A clinical study of root surface conditioning with an EDTA gel. I. Non surgical periodontal treatment. *The International Journal of Periodontics and Restorative Dentistry*, *20*, 560-565.

[49] Mariotti, A. (2003). Efficacy of chemical root surface modifiers in the treatment of periodontal disease. A systematic review. *Annals of Periodontology*, *8*, 205-226.

[50] Hammarström, L., Heijl, L. & Gestrelius, S. (1997). Periodontal regeneration in a buccal dehiscence model in monkey after application of EMD. *Journal of Clinical Periodontology*, *249*, 669-677.

[51] Yukna, R. A. & Mellonig, J. T. (2000). Histologic evaluation of periodontal healing in humans following regenerative therapy with enamel matrix derivative. *Journal of Periodontology*, *71*, 752-759.

[52] Sculean, A., Donos, N., Windisch, P., Brecx, M., Gera, I., Reich, E. & Karring, T. (1999). Healing of human intrabony defects following treatment with enamel matrix proteins or guided tissue regeneration. *Journal of Periodontal Research*, *34*, 310-322.

[53] Donos, N., Sculean, A., Glavind, L., Reich, E. & Karring, T. (2003). Wound healing of degree III furcation involvements following guided tissue regeneration and/or Emdogain. A histologic study. *Journal of Clinical Periodontology*, *30*, 1061-1068.

[54] Cochran, D. L., King, G. N., Schoolfield, J., Velasquez-Plata, D., Mellonig, J. T. & Jones, A. (2003). The effect of enamel matrix proteins on periodontal regeneration as determined by histological analyses. *Journal of Periodontology*, *74*, 1043-1055.

[55] Koop, R., Merheb, J. & Quirynen, M. (2012). Periodontal regeneration with enamel matrix derivative in reconstructive periodontal therapy: a systematic review. *Journal of Periodontology*, *83*, 707-720.

[56] Jepsen, S., Heinz, B., Jepsen, K., Arjomand, M., Hoffmann, T., Richter, S., Reich, E., Sculean, A., Gonzales, J. R., Bödeker, R. H. & Meyle, J. (2004). A randomized clinical trial comparing enamel matrix derivative

and membrane treatment of buccal Class II furcation involvement in mandibular molars. Part I: Study design and results for primary outcomes. *Journal of Periodontology*, *75*, 1150-1160.

In: EDTA
Editor: Andris Molnar
ISBN: 978-1-62808-146-6

Chapter 5

EFFECTS OF EDTA ON ADSORPTION BEHAVIOR OF CADMIUM ONTO NATURAL FRESHWATER SURFACE COATINGS

Xiuyi Hua**, Deming Dong, Dapeng Liang and Zhiyong Guo

Key Lab of Groundwater Resources and Environment of Ministry of Education, Key Lab of Water Resources and Aquatic Environment of Jilin Province, College of Environment and Resources, Jilin University, Jilin, China

ABSTRACT

Natural surface coatings play important roles in determining behavior of pollutants in aquatic ecosystems. Natural surface coatings usually possess remarkable heavy metal adsorption abilities. EDTA is one of the common artificial ligands encountered in natural waters. It is also a commonly used surrogate for natural organic matters, to assess the roles of natural organic matters in natural aquatic environments. In this study, the effects of EDTA on adsorption behavior of cadmium onto natural freshwater surface coatings were investigated, including adsorption kinetics, adsorption isotherm and pH-edge adsorption. The influences of solution pH and EDTA concentration were also were investigated. The results indicated that EDTA usually has considerable effects on adsorption of

* Corresponding author: e-mail address: huaxy@jlu.edu.cn; Tel./fax.: 86-431-85168429.

cadmium to natural surface coatings. EDTA made the adsorption process approach equilibrium more quickly. EDTA decreased the adsorption of cadmium to surface coatings, especially when the ratio of concentration of EDTA to that of cadmium was high, or when the pH of adsorption solutions was high. The dissolution of some surface coating components was also observed. These effects of EDTA could be explained in terms of the competition between EDTA and the surface coatings for available cadmium, the converting of positively charged cadmium ions to neutrally or even negatively charged species, and the dissolution of manganese oxides from the surface coatings.

INTRODUCTION

To investigate the fate and transport of trace metals, the effects of major controlling physical, chemical and biological processes should be assessed. Adsorption to the different solid surfaces is one of the key processes controlling metal ion mobility in aquatic environment, and a number of factors affect the adsorption behavior of metals including pH, organic ligands and competing sorbates [1-3]. It has been found that organic ligands may either enhance or retard the migration of metals in natural water systems, depending on the system composition [1, 4, 5]. Dissolved organic matters (DOM) exist widely in natural aquatic ecosystems. As they usually act as the ligands, their presence affects the fate of trace metals in water significantly, by changing the speciation of trace metals and the adsorption/desorption behavior of trace metals [4-9]. In natural waters, DOM comprise different types of organic matters with different structure and origin [4, 5, 7]. Therefore, the effects of DOM on trace metals are usually investigated using a certain type of DOM or some simple surrogates, including some polymeric DOM such as humic substances [10-12] and synthetic polyelectrolytes [9, 13], and/or some small organic molecules such as EDTA, oxalic acid and citric acid, etc. [8, 14-16]. The effects of DOM on metal adsorption are also complex, and both enhancement and diminishment effects have been observed [4, 17, 18]. Furthermore, the roles of DOM in trace metal adsorption are affected by such factors as the type and the concentration of the involved adsorbents, ligands and metals [8, 18-20], the solution pH [4, 9, 11, 17], the coexisting components [11, 21, 22], and the adsorption sequence [9, 12, 14, 21], etc. The mechanisms of the effects of DOM on the adsorption are also very complicated [4, 23-26].

In aquatic environments, natural surface coatings also exist widely. These sort of heterogeneous solid materials consist of microorganisms and associated minerals, and occur on surfaces of all kinds of solid matters in natural waters [27-29]. They have strong affinities for trace pollutants, adsorb and scavenge pollutants from the water, and change the behavior and fate of the pollutants significantly [28-32]. The interaction between surface coatings and pollutants has attracted great interest in recent years [29-33]. Studies have been carried out to investigate the adsorption of trace metals to surface coatings and their main components [34-39], but the effects of DOM on the adsorption are still not clear.

In natural aquatic environments, the coexisting DOM affect the adsorption of trace metals to the surface coatings. However, details of the effects are still not clearly understood. Therefore, investigations on the effects of DOM on trace metal adsorption to surface coatings are critically needed. In this study, the effects of EDTA, a representative component of DOM, on the adsorption of cadmium, a typical trace metal in natural water, to natural surface coatings, including adsorption kinetics, adsorption isotherms and pH-edge adsorption, were examined. The influences of solution pH and relative concentration of EDTA on the effects of EDTA were also evaluated. EDTA is not only a frequently used simple representative organic ligand in studying the role and behavior of DOM [4, 14, 23, 24, 40], but also a chemical used extensively in industries (metal processing, photographic industry and industrial cleaning, etc.) [41-43]. In addition, it is a commonly used chelating agent for the remediation of contaminated soils [44, 45]. In recent years, EDTA has been found to occur at significant concentrations in natural waters, due to its low biodegradability [42, 43, 46]. This study would give us a better understanding of the behavior and fate of trace metals in natural waters.

MATERIALS AND METHODS

Development of Natural Surface Coatings

Nanhu Lake, a urban eutrophic lake in Changchun, China, was selected as the body of natural water for the development of natural surface coatings [36, 38, 47]. Surface coatings were developed on glass slides (48 mm × 76 mm × 1 mm) held in polypropylene racks, which were submerged in the lake water at a depth of about 30 cm for 10 days [34-38]. After development, glass slides with attached surface coatings were transported to laboratory (submerged in lake water) for characterization and cadmium adsorption. As important components to the adsorption of trace metals [34, 35, 48], surface concentrations of the Fe and Mn oxides and organic material in the surface coatings were measured. Total acid-extractable Fe and Mn oxides in the surface coatings were determined by extracting the slides of surface coatings with 15% HNO_3 for 24 h [35, 36, 49], and Fe and Mn in the acid extracts were analyzed by flame or graphite furnace atomic absorption spectrometry (FAAS/GFAAS) using a Perkin Elmer AAnalyst 700 equipped with a HGA850 graphite furnace. Organic material in the surface coatings was quantified by measuring total organic carbon (TOC) using a Shimadzu TOC-V_{CPH} total organic carbon analyzer equipped with an SSM-5000A solid sample combustion unit.

Adsorption Experiments

The adsorption experiments were conducted in a minimal mineral salts (MMS) solution in laboratory. The MMS solution contained 200 μmol/L $CaCl_2$, 140 μmol/L $MgSO_4$, 910 μmol/L $(NH_4)_2SO_4$, 150 μmol/L KNO_3 and 10 μmol/L $NaHCO_3$, and was used to simulate natural fresh water and to maintain the biological activity of the surface coatings [34-36]. EDTA was prepared by its disodium salt and cadmium was prepared by $Cd(NO_3)_2$. Three types of experiments were carried out for the adsorption in this study: adsorption kinetics at a fixed pH (pH = 6.0), with the absence and presence of EDTA, respectively; isotherm adsorption at a pH = 6.0, with the absence and presence of EDTA; pH-edge adsorption with two EDTA concentrations. Speciation of cadmium in the MMS solutions was calculated with Visual MINTEQ (version 2.13).

Adsorption Kinetics at pH = 6.0

Adsorption kinetics of cadmium to natural surface coatings was measured at pH = 6.0 and [Cd] = 0.5 μmol/L, without EDTA and with the presence of EDTA (0.5 μmol/L), respectively. Two slides of surface coatings were placed in polypropylene racks and submerged into each of the four 800-mL MMS solutions with cadmium and with or without EDTA. Two of the solutions were duplicates for adsorption without EDTA and other two were duplicates for adsorption with EDTA. The solutions were contained in 2-L water-jacketed beakers to maintain a constant temperature of 25 ± 1 °C. The solutions were stirred continuously with magnetic stirrers while maintaining the pH at 6.0 ±

0.1 using pH controllers (Dejiali Scientific Inc., Qingdao, China) to regulate the addition of 0.01 M HNO_3 and NaOH. The adsorption was performed for 600 min. These controlled adsorption conditions were applied to eliminate the effects of factors other than EDTA on the adsorption and also to make this study comparable to other studies [34-38]. 3-ml solutions were sampled from each of the beakers at different time intervals, and totally 15 samples were collected from each of the beaker. Cadmium in the solutions collected was analyzed by GFAAS. The amount of cadmium adsorbed to the surface coatings at each sampling time was calculated as the difference between the initial concentration and the remaining concentration of cadmium in the solutions.

Adsorption kinetics data were fitted into the pseudo-second-order adsorption kinetics model, in its Langmuir isotherm like form (Langmuir kinetics model) [50-52]:

$$\Gamma_t = \frac{\Gamma_{eq} kt}{1 + kt}, \tag{1}$$

where Γ_t is the adsorption of cadmium by surface coatings (μmol/m^2, concentrations of cadmium or components in the surface coatings are expressed as per unit nominal surface area of the glass slides in this study) at time t, Γ_{eq} is the equilibrium cadmium adsorption (μmol/m^2), k is a coefficient whose reciprocal is the time needed when half amount of the equilibrium cadmium adsorption is adsorbed (1/min). The model was fitted using nonlinear least squares method by Microcal Origin (version 6.1, Microcal Software, Inc.)

Adsorption Isotherm at pH = 6.0

Adsorption isotherms were obtained by measuring cadmium adsorption from solutions at six different cadmium concentrations ranging from 0.3 to 2.3 μmol/L, without EDTA and with EDTA (0.5 μmol/L), respectively. Two slides of surface coatings were placed into each of the twelve 800-mL MMS solutions with different cadmium and EDTA concentrations. The adsorption was performed for 24 h at 25 ± 1 ºC and pH = 6.0 ± 0.1. After equilibration, each slide was extracted into 25 mL of 15% HNO_3 for 24 h. cadmium in the acid extracts and the equilibrium solutions were also analyzed by FAAS or GFAAS.

Adsorption data of cadmium were fitted into Langmuir isotherm [34, 38]:

$$\Gamma = \frac{\Gamma_{max} K[\mathrm{Cd}]}{1 + K[\mathrm{Cd}]}, \quad (2)$$

where Γ is the adsorption of cadmium by surface coatings (μmol/m^2, concentrations of cadmium or components in the surface coatings are expressed as per unit nominal surface area of the glass slides in this study), Γ_{max} is the maximum cadmium adsorption (μmol/m^2), K is the Langmuir equilibrium coefficient (L/μmol), and [Cd] is the equilibrium concentration of cadmium (μmol/L).

PH-edge Adsorption

Two sets of pH-edge adsorption experiments were conducted: cadmium only adsorption and cadmium adsorption with EDTA. The concentration of cadmium in the solutions was 2.0 μmol/L. For EDTA, two concentrations were selected, 0.5 and 2.0 μmol/L. These adsorption experiments were carried out at 12 different pH values ranging from 4.3 to 9.0 for 24 h. The pH was maintained during the period of adsorption, also using automatic pH controllers. After equilibrium, the final pH values were measured as the adsorption pH of the corresponding beakers. Other details were the same as those described for kinetics and isotherm adsorption.

RESULTS AND DISCUSSION

Surface coatings developed in Nanhu Lake consisted of microorganisms and associated mineral deposits. The surface coatings were brownish green in color and contained large numbers of diatoms, green algae, cyanobacteria and bacterial cells. The average surface concentrations of the main components of the surface coatings were 106.3 ± 9.0 μmol Fe/m^2 (n = 10), 141.5 ± 8.5 μmol Mn/m^2 (n = 10) and 163.7 ± 6.4 mg TOC/m^2 (n = 4) for acid-extractable Fe oxides and Mn oxides, and organic material, respectively. Therefore, the surface coatings were consistent from slide to slide (component concentrations varied by less than 9%), allowing the use of different slides for the measurement of cadmium adsorption.

Adsorption Kinetics at the Fixed pH

Adsorption process of cadmium to natural surface coatings at pH = 6.0, without and with the presence of EDTA, is shown in Figure 1. The adsorption process, with or without EDTA, followed the pseudo-second-order adsorption kinetics model (the Langmuir kinetics model). The fitted parameters were: $\Gamma_{eq} = 4.64 \pm 0.06$ μmol/m^2, $k = 0.0157 \pm 0.0007$ 1/min and R (correlation coefficient of the fitting) = 0.9984 for cadmium only adsorption; and $\Gamma_{eq} = 2.02 \pm 0.02$ μmol/m^2, $k = 0.0252 \pm 0.0012$ 1/min and $R = 0.9983$ for cadmium adsorption with EDTA. At this concentration ratio, [Cd] : [EDTA] = 1 : 1, the presence of EDTA decreased the cadmium adsorption significantly to less than half, which will be discussed extensively in the "Adsorption Isotherm at the Fixed pH" section. From the fitted parameter k, the presence of EDTA decreased the time needed when half of the total equilibrium adsorption is finished, and therefore made the adsorption approach equilibrium more quickly. This can be seen more clearly form the ratio of cadmium adsorption in system with EDTA to that in system without EDTA (Figure 2 (A)) and the comparison of the normalized adsorption kinetics processes (Figure 2 (B)). Here, the normalization refers to divide the kinetics equation (Eq. (1)) by the equilibrium adsorption (Γ_{eq}), to obtain a kinetics equation about Γ_t/Γ_{eq} (Eq. (3)), which facilitates the comparison of adsorption process between the adsorption kinetics with different Γ_{eq} values:

$$\frac{\Gamma_t}{\Gamma_{eq}} = \frac{kt}{1+kt}, \tag{3}$$

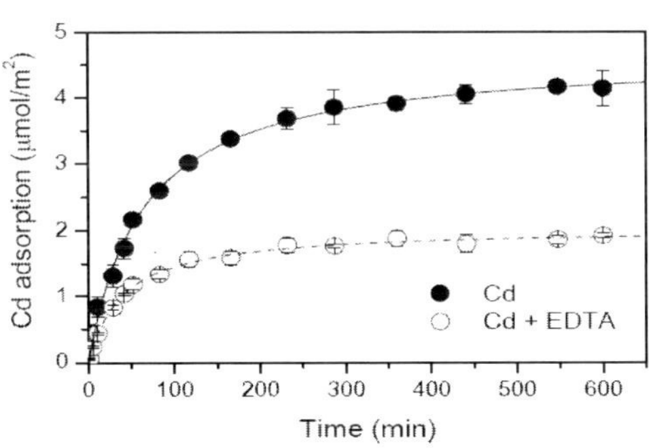

Figure 1. Adsorption process of cadmium to natural surface coatings at pH = 6.0 in the absence and presence of EDTA at 0.5 μmol/L. Error bars indicate ± one deviation (n = 2).

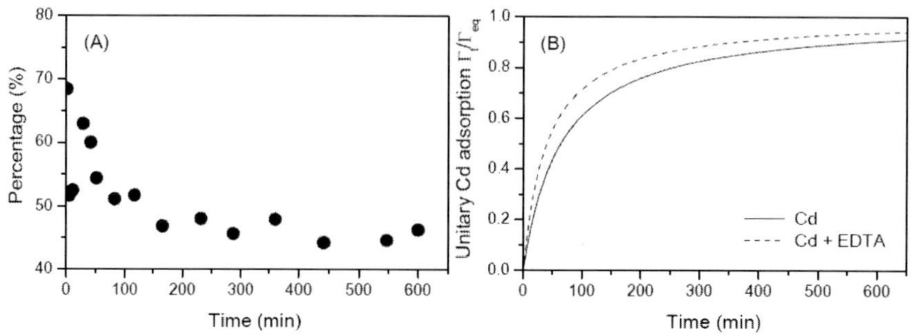

Figure 2. The effect of EDTA at 0.5 μmol/L on cadmium adsorption kinetics. (A) The percentage of the amount of adsorbed cadmium in system with EDTA to that in system without EDTA; (B) the normalized adsorption kinetics of cadmium to natural surface coatings with and without EDTA.

This effect of EDTA should be due to that the presence of EDTA decreased the adsorption. It has been found that quicker adsorption usually associated with less metal adsorption [27].

Adsorption Isotherm at the Fixed pH

The Effect of EDTA on Adsorption Isotherm

Cadmium adsorption to natural surface coatings at pH = 6.0, without and with the presence of EDTA, is shown in Figure 3. Without EDTA, the adsorption followed the Langmuir isotherm. By fitting the adsorption data into Langmuir isotherm, the following parameters could be obtained: $\Gamma_{max} = 17.6 \pm 1.2$ μmol/m^2, $K = 0.80 \pm 0.11$ L/μmol and $R = 0.9975$. With the presence of EDTA (0.5 μmol/L), the adsorption decreased significantly, especially when the equilibrium cadmium concentration was low, which is consistent with the variation of Γ_{eq} observed in the "Adsorption Kinetics at the Fixed pH" section. This made the adsorption isotherm of cadmium to surface coatings with the presence of EDTA deviated from the Langmuir isotherm (Figure 3). Therefore, the data were not fitted into the Langmuir isotherm.

It has been found that EDTA usually decreases the adsorption of trace metals [14, 21, 41]. To quantify this effect, we calculated the percentages of the adsorbed cadmium after equilibrium to the total amount of cadmium added to each of the system, at each corresponding initial cadmium concentrations, in the adsorption systems without and with the presence of EDTA (Figure 4). After equilibrium, about 20% or less of the cadmium was adsorbed, in adsorption systems without or with the presence of EDTA. In systems without EDTA, the fraction of the adsorbed cadmium decreased with the increase of initial cadmium concentration, as determined by the Langmuir isotherm. While in systems with EDTA, the fraction of the adsorbed cadmium was lower than the corresponding fraction in system without EDTA, especially when initial cadmium concentration was low. With the increase of initial cadmium concentration, the fraction of the adsorbed cadmium increased first, and then remain almost constant, which disagreed with the trend of the Langmuir isotherm, until the initial concentration was high (> 1.7μmol/L). We also estimated the adsorption of cadmium without EDTA at the equilibrium cadmium concentrations exactly the same to those of the adsorption with EDTA, according to the fitted Langmuir isotherm of the adsorption without EDTA. Based on the calculated adsorption data for adsorption without EDTA and the measured data for adsorption with EDTA, at the same equilibrium concentrations, the amounts of the reduced cadmium adsorption due to the presence of EDTA could be calculated, and so did the percentages of these decrements of adsorption to the calculated adsorption without EDTA, which are shown in Figure 4. The percentage of decreased cadmium adsorption decreased dramatically with the increase of equilibrium cadmium concentration, from as high as 84% to about 1%. This indicted the effect of EDTA on cadmium adsorption was closely related to the ratio of [Cd]/[EDTA]. The smaller the ratio was, the more remarkable the relative effect would be.

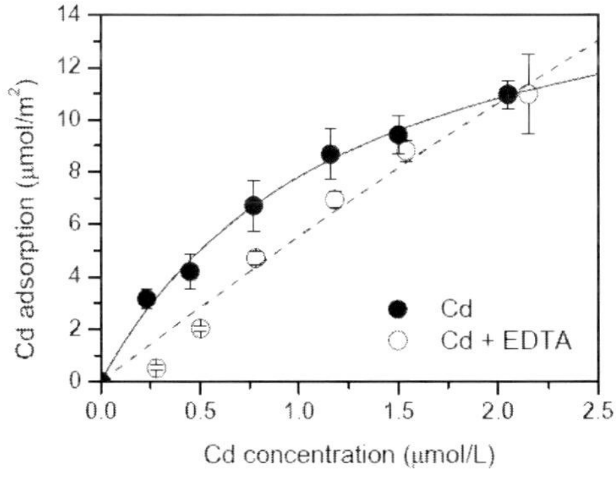

Figure 3. Adsorption isotherms of cadmium to natural surface coatings at pH = 6.0 in the absence and presence of EDTA at 0.5 μmol/L. Error bars indicate ± one deviation (n = 2).

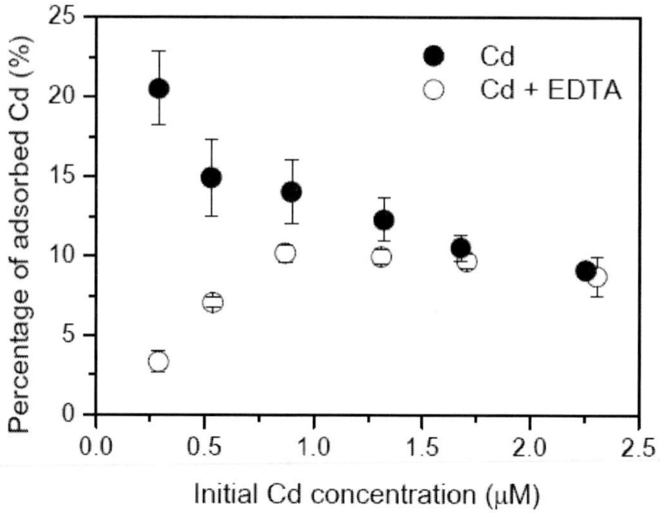

Figure 4. Adsorption isotherms of cadmium to natural surface coatings at pH = 6.0 in the absence and presence of EDTA at 0.5 μmol/L. Error bars indicate ± one deviation (n = 2).

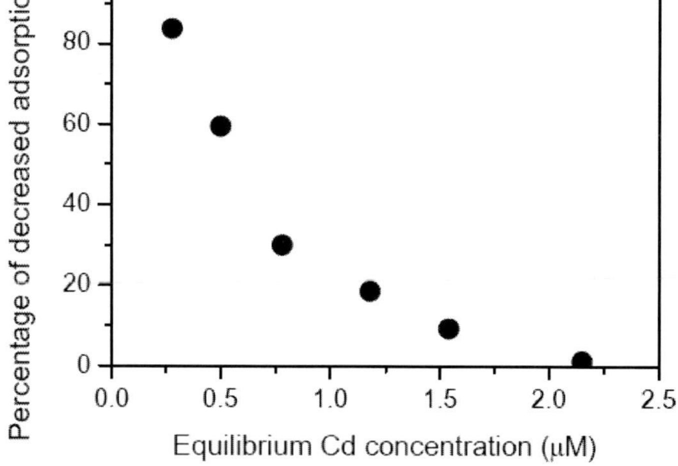

Figure 5 Percentage of the decreased cadmium adsorption in systems with EDTA (0.5 μmol/L), to the total cadmium adsorption in systems without EDTA at corresponding equilibrium concentrations, which were calculated according to the fitted isotherm.

The Effect of EDTA on the Dissolution of Metal Oxides from the Surface Coatings

It has been suggested that EDTA might dissolve some solid materials, such as metal oxides, in aquatic environments [53]. If EDTA has dissolved some of the metal oxides from the surface coatings, the adsorption would definitely decrease. Therefore, the contents of Fe and Mn oxides in all the surface coatings after the adsorption without or with EDTA were all measured. The average surface concentration of Fe oxides was 107.6 ± 8.4 μmol Fe/m^2 (n = 12) for surface coatings after adsorption without EDTA, and was 105.3 ± 5.4 μmol Fe/m^2 (n = 12) for adsorption with EDTA. While for Mn oxides, the concentrations were 75.2 ± 6.5 and 78.3 ± 4.9 μmol Mn/m^2 (n = 12), respectively. Therefore, at least under the experiment conditions (pH = 6.0, [EDTA] = 0.5 μmol/L), the EDTA had little significant effect on the contents of Fe or Mn oxides in the surface coatings, during the period of adsorption. However, by comparing the concentrations of metal oxides after adsorption with those determined before adsorption, it could be found that the adsorption process did decrease the concentration of Mn oxides in the surface coatings considerably. The phenomenon has also been observed [27, 36, 38]. This dissolution of Mn oxides should be induced by the relatively low pH of the adsorption solution, as the surface coatings were developed in the lake water whose pH was around 8.5 (the average pH of the lake water during the period of development) while the adsorption was carried out in the MMS solution whose pH was fixed at 6.

The Effect of EDTA on Speciation of Cadmium in Solutions

As EDTA is a strong ligand for trace metals, the speciation of cadmium in the MMS solution should be affected severely by the EDTA when it was added into the solution. This converting of speciation would subsequently affect the adsorption behavior of cadmium significantly [1-3]. Due to the complexity of the surface coatings in the solutions, the discussion was based on the calculated speciation. The calculated speciation of cadmium in the MMS solutions at pH = 6.0 is shown in Figure 6. From Figure 6(A), without EDTA, the fraction of each cadmium species did not change with the variation of cadmium concentration, and was dominated by Cd^{2+} (85.5%). From Figure 6(B), when total cadmium concentration was less than 0.5 μmol/L, almost all of the cadmium combined with EDTA to form $CdEDTA^{2-}$. When total cadmium concentration was greater than 0.5 μmol/L, the fraction of $CdEDTA^{2-}$ decreased with the increase of total cadmium concentration, while the fractions of other species increased correspondingly.

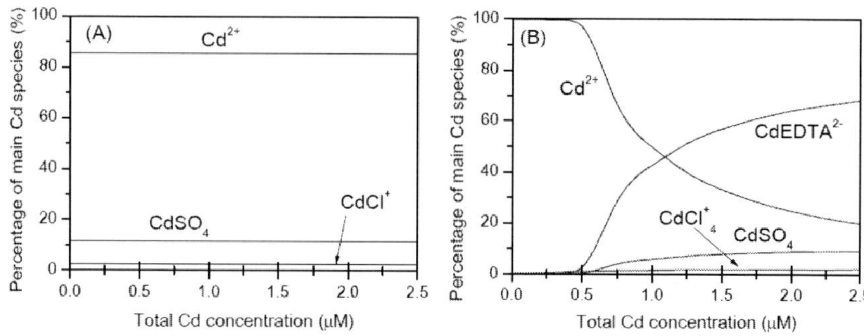

Figure 6. Speciation of cadmium in the MMS solutions at different cadmium concentration (pH = 6.0). Only main species (fraction > 1%) are shown. (A): Speciation of cadmium without EDTA; (B): Speciation of cadmium with EDTA, at [EDTA] = 0.5 μmol/L.

Mechanism of the Effect of EDTA on Adsorption Isotherm

By comparing Figure 6(A) with Figure 6(B), it could be concluded that both cadmium and EDTA tend to combine with each other at an equivalent molar ratio, to form the negatively charged complex, $CdEDTA^{2-}$, which decreased the available Cd^{2+} and resulted in the decreased cadmium adsorption. The higher the fraction of $CdEDTA^{2-}$ became, the more considerable the decrease would be. In Figure 6(B), at lower cadmium concentration range ([Cd] < 0.6 μmol/L), the fraction of free Cd^{2+} was very low (approaching 0), but the fraction of the adsorbed cadmium was not so low. This indicated the contribution of $CdEDTA^{2-}$ to the adsorption of cadmium, though the contribution was relatively low. There are at least two main mechanisms for the adsorption of cadmium in the Cd-EDTA complexes to the solid surface: 1) the surface of the surface coatings "captured" some of the cadmium from the Cd-EDTA complexes in the solution, like a ligand-exchange reaction [1, 21, 42]:

$$\equiv S^{m-} + Cd\text{-}L^{(n-2)-} \rightleftharpoons \equiv S\text{-}Cd^{(m-2)-} + L^{n-}, \tag{4}$$

here, $\equiv S^{m-}$ represents the adsorption site on surface of the surface coatings, L^{n-} represents species of EDTA such as $EDTA^{4-}$ or $HEDTA^{3-}$, etc.; 2) the direct adsorption of $CdEDTA^{2-}$ onto the surface through the formation of outer sphere ternary surface complexes (a ≡S-(Cd-EDTA) structure) and/or inner sphere A-type ternary surface complexes (a ≡S-Cd-EDTA structure) [23, 24,

40, 42]. Therefore, under the experimental conditions, as the surface of solid materials was usually negatively charged [1-3], the adsorption of cadmium to the surface coatings was mainly contributed by Cd^{2+} or other positively charged species, and the contribution of $CdEDTA^{2-}$ was limited.

PH-dependent Adsorption

The Effect of EDTA on PH-dependent Adsorption

The adsorption of cadmium to surface coatings at different pH is shown in Figure 7. Without or without different levels of EDTA, the adsorption of cadmium all presented the typical "S-shaped" metal adsorption curves [1-3, 13]. As shown in the figure, the presence of EDTA changed the adsorption behavior of cadmium, and the effect was influenced by both the solution pH and the concentration of EDTA. When [EDTA] = 0.5 μmol/L, if pH was less than 7.0, the presence of EDTA had no significant effect on the adsorption; while if pH was grater than 7.5, EDTA decreased the adsorption (up to about 25%). While when [EDTA] = 2.0 μmol/L, the decrease of adsorption was very pronounced (more than 60% when pH was grater than 4.5, and up to about 90%).

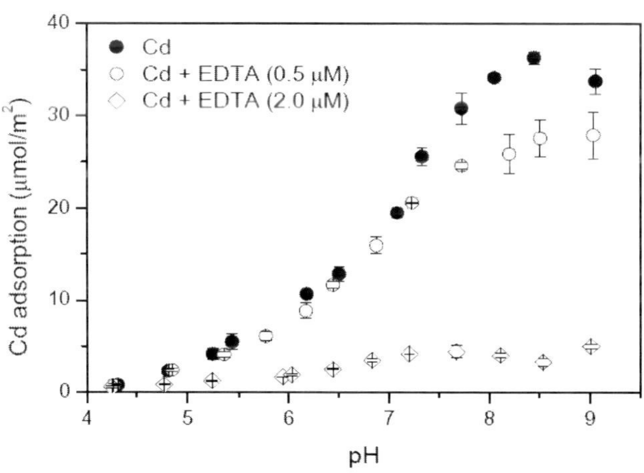

Figure 7. pH adsorption-edge of cadmium to natural surface coatings, including cadmium only adsorption and adsorption in the presence of EDTA at two concentrations (0.5 μmol/L and 2.0 μmol/L). Error bars indicate ± one deviation (n = 2).

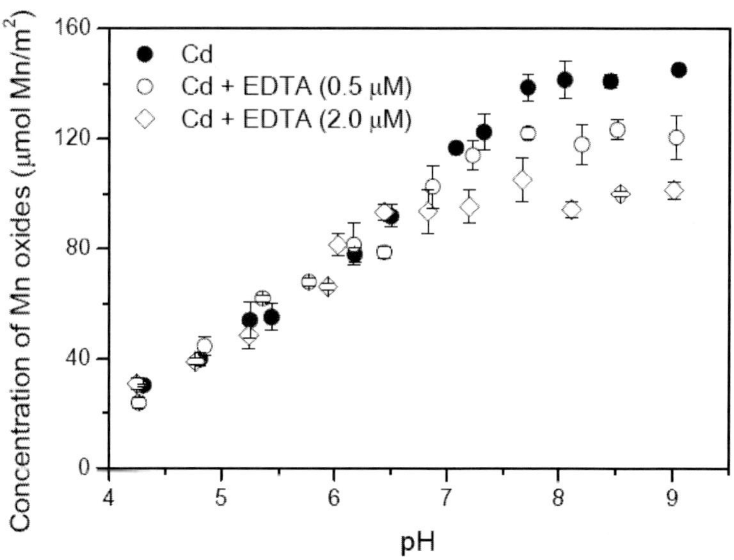

Figure 8. Concentration of Mn oxides in natural surface coatings after the cadmium adsorption at different pH, including cadmium only adsorption and adsorption in the presence of EDTA at two concentrations (0.5 μmol/L and 2.0 μmol/L). Error bars indicate ± one deviation (n = 2).

The Effect of EDTA on the Dissolution of Metal Oxides from the Surface Coatings at Different pH

The effects of EDTA on the adsorption at different pH were also discussed from the dissolution of components of the surface coatings and from the transformation of dissolved cadmium species. Determination of the metal oxides remaining in the surface coatings after the adsorption process indicated that the presence of EDTA and the relatively low pH of the adsorption solutions had no appreciable effects on the contents of Fe oxides in all the surface coatings, again. Concentrations of Fe oxides in these surface coatings were still consistent from slide to slide, and the average concentration for all of the surface coatings was 104.4 ± 7.5 μmol Fe/m^2 (n = 72), almost the same to the concentration determined before the adsorption. However, concentrations of Mn oxides were critically affected by the EDTA and the pH of the solutions, as shown in Figure 8. When pH was less than 7.5, the effect of pH on the dissolution was extremely remarkable, and the concentrations of Mn oxides decreased almost linearly with the decrease of pH. For example, for adsorption without EDTA, the concentration of Mn oxides decreased from about 138 μmol Mn/m^2 at pH = 7.7, which was similar to the initial Mn concentration, to only about 30 μmol Mn/m^2 at pH = 4.3. The effect of EDTA was insignificant when pH was less than 6.2, and was very remarkable when pH was greater than 7.5. It seems that the dissolution effect of EDTA on Mn oxides in the surface coatings was complementary to that effect of pH: the EDTA played a role in the higher pH range while the effect of pH was more considerable in the relatively lower pH range. The effect of EDTA was also determined by its concentration: the dissolution at [EDTA] = 2.0 μmol/L was greater than that at [EDTA] = 0.5 μmol/L.

Here, Mn oxides in the natural surface coatings were found to be very sensitive to pH and EDTA. However, some researchers found that the dissolution effect of EDTA on Mn oxides in soils and sediments was usually insignificant, when compared with Fe oxides and other minerals [54-56]. Biogenic Mn oxides and fresh abiotic Mn oxides have been found to be dissolved much easier [57-59]. Therefore, this confirmed that unlike Mn oxides in soils and sediments, Mn oxides in natural surface coatings were usually dominated by biogenic or fresh ones, whose reactivity were usually high [34-36]. Furthermore, the formation of Mn oxides has been found to occur frequently in natural water [2, 34, 58], and Mn oxides in natural surface coatings usually possess high specific adsorption capability to trace metals as well [34-36, 39, 48]. Therefore, as a highly active and variable component with remarkable trace metal adsorption ability, the role of Mn oxides in surface coatings in affecting the behavior of trace metals might be more significant than we have expected. The results of this study also imply that Fe oxides in surface coatings should be fairly stable. Indeed, higher EDTA concentration might be needed to dissolve the Fe oxides [53].

The Effect of EDTA on Speciation of Cadmium in Solutions at Different pH

The fractions of species of cadmium in MMS solutions at different pH values were plotted in Figure 9. From Figure 9 (A), without EDTA, cadmium was dominated by Cd^{2+} when pH was less than 8.5. When pH was greater than 8.0, other species, especially $CdCO_3(aq)$ and $Cd(CO_3)_2^{2-}$, increased and became major species with the increase of pH. With the presence of EDTA at [EDTA] = 0.5 μmol/L (Figure 9 (B)), beside the complexes formed with EDTA (mainly $CdEDTA^{2-}$), the speciation of cadmium was almost the same to that of cadmium without EDTA (Figure 9 (A)), as the concentration of cadmium was grater than that of EDTA. With [EDTA] = 2.0 μmol/L (Figure 9 (C)), almost all the cadmium combined with EDTA cadmium to form the complexes, also mainly $CdEDTA^{2-}$, in the whole pH ranges. As indicated above, the surface of the surface coatings is usually negatively charged. Therefore, the adsorption of negatively charged cadmium species should be much more difficult than the adsorption of positively charged cadmium species and neutral cadmium species. The total percentages of the positively charged cadmium species and neutral cadmium species in the MMS solutions at different pH, which were calculated form the speciation of cadmium shown in Figure 9, are illustrated in Figure 10. From the figure, EDTA decreased the percentage of the positively charged cadmium species and neutral species significantly. High pH also decreased the percentage of the positively charged species (pH > 8.0) and non- negatively charged species (pH > 8.5).

Mechanism of the Effect of EDTA on PH-dependent Adsorption

The "S-shaped" adsorption have been reported by many researchers for the adsorption of positively charged metal ions onto different types of solid matters without the presence of EDTA, and have been attributed to the effects of pH on the protonation and deprotonation of metal-binding functional groups on the solid matters, on the competition between H^+ and trace metal ions, and on the hydrolysis and precipitation of trace metals, etc. [1-3, 13]. As to natural surface coatings in this study, besides the above well-known mechanisms, the dissolution of the Mn oxides in the surface coatings was also an important reason.

From the above results, the effects of EDTA on the adsorption of cadmium to surface coatings could be attributed to the complexation of EDTA with cadmium, the change of surface adsorption reactions, and the dissolution of the Mn oxides from the surface coatings. Furthermore, these effects were also influenced by the concentration of EDTA (more accurately, the [EDTA]/[Cd]) and the solution pH. As to the concentration of EDTA, it is obvious that when the concentration was low (relative to [Cd]), only some of the cadmium was complexed and only less amount of Mn oxides was dissolved. For pH, as EDTA is a weak acid, higher pH facilitates its dissociation and its complexation with metal ions [1, 60]. Therefore, the ability of EDTA to compete with surface coatings for the available cadmium to form complexes increased with the increase of pH. So did the ability to dissolve Mn oxides from the surface coatings (to form soluble complexes of Mn-EDTA). Besides, the adsorption of metal-EDTA complexes is a "ligand-like" adsorption, which decreases dramatically with the increase of pH [1, 42]. Consequently, the effect of EDTA was more pronounced in the higher pH range, especially when the ratio of [EDTA]/[Cd] was relatively low (Figure 7).

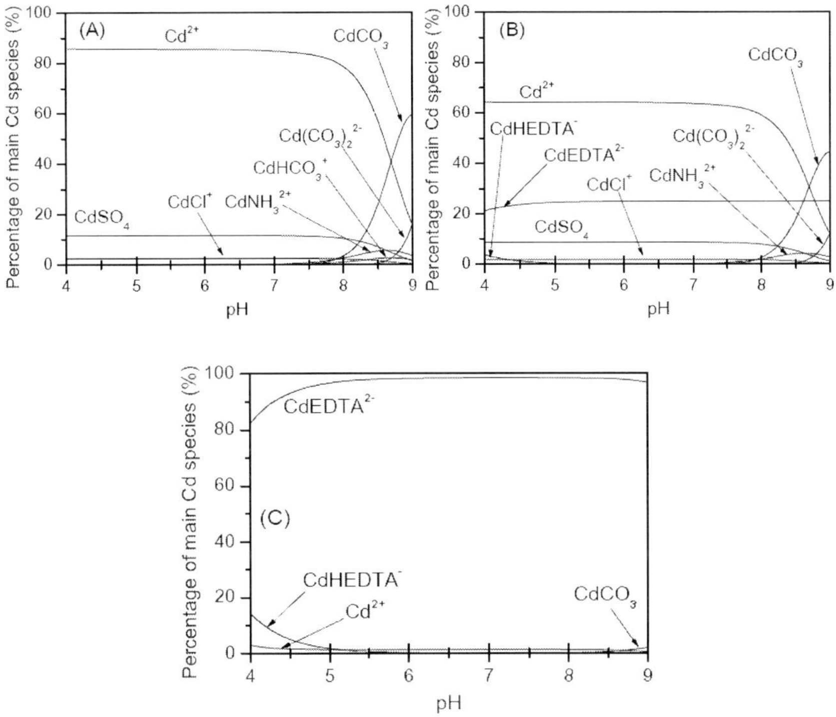

Figure 9. Speciation of cadmium ([Cd] = 2.0 μmol/L) in the MMS solutions at different pH. Only main species (fraction > 1%) are shown.
(A): [EDTA] = 0 μmol/L; (B): [EDTA] =0.5 μmol/L; (C): [EDTA] = 2.0 μmol/L.

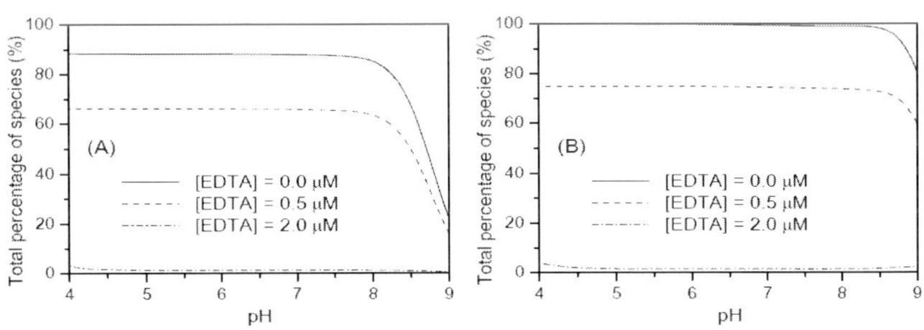

Figure 10. Total percentage of positively charged cadmium species and neutral cadmium species ([Cd] = 2.0 μmol/L) in the MMS solutions at different pH.
(A): positively charged cadmium species; (B): Both positively charged and neutral species of cadmium.

CONCLUSION

The presence of EDTA generally decreased the adsorption of cadmium to natural surface coatings and made the adsorption processes approach equilibrium more quickly. The effects were determined by the concentration of EDTA and the solution pH. When the concentration of EDTA was similar to or higher than that of cadmium, or when the solution pH was relatively higher, the effect of EDTA on adsorption was more significant. The effect of EDTA on cadmium adsorption to surface coatings was mainly due to the conversion of cadmium species to negatively charged complexes, the competition of EDTA with the surface of surface coatings for the dissolved free cadmium, and the dissolution of Mn oxides from the surface coatings.

ACKNOWLEDGMENTS

This study was supported by the National Natural Science Foundation of China (No. 20607007 and 20877033) and the Major Science and Technology Program for Water Pollution Control and Treatment in China (No. 2012ZX07202009).

REFERENCES

[1] Stumm, W. & Morgan, J. J. (1996). *Aquatic Chemistry* (3rd ed.); Wiley: New York.

[2] Warren, L. A. & Haack, E. A. (2001). *Earth Sci. Rev.*, *54*, 261-320.

[3] Sposito, G. (2004). *The Surface Chemistry of Natural Particles*; Oxford University Press: New York.

[4] Kantar, C. (2007). *Earth Sci. Rev.*, *81*, 175-198.

[5] Frimmel, F. H. (1998). *J. Contam. Hydrol.*, *35*, 201-216.

[6] Leenheer, J. A. & Croué, J. P. (2003). *Environ. Sci. Technol.*, *37*, 18A-26A.

[7] Merdy, P., Koopal, L. K. & Huclier, S. (2006). *Environ. Sci. Technol.*, *40*, 7459-7466.

[8] Minaberry, Y. S. & Gordillo, G. J. (2010). *Chemosphere*, *78*, 1356-1361.

[9] Hua, X., Dong, D., Jiang, X., Hu, J., Guo, Z. & Liang, D. (2012). *Chem. J. Chin. Univ.*, *33*(9), 2005-2012.

[10] Christl, I., Metzger, A., Heidmann, I. & Kretzschmar, R. (2005). *Environ. Sci. Technol.*, *39*, 5319-5326.

[11] Chen, C. & Wang, X. (2007). *Appl. Geochem.*, *22*, 436-445.

[12] Xu, D., Zhou, X. & Wang, X. (2008). *Appl. Clay Sci.*, *39*, 133-141.

[13] Floroiu, R., Davis, A. & Alba, T. (2001). *Environ. Sci. Technol.*, *35*, 348-353.

[14] Bryce, A. L., Kornicker, W. A. & Elzerman, A. W. (1994). *Environ. Sci. Technol.*, *28*, 2353-2359.

[15] Huang, L., Hu, H., Li, X. & Li, L. (2010). *Appl. Clay Sci.*, *49*, 281-287.

[16] Li, X., Wu, J., Liao, J., Zhang, D., Yang, J., Feng, Y., Zeng, J., Wen, W., Yang, Y., Tang, J. & Liu, N. (2013). *J. Environ. Radioact.*, *115*, 143-150.

[17] Lackovic, K., Wells, J. D., Johnson, B. B. & Angove, M. J. (2004). *J. Colloid Interface Sci.*, *270*, 86-93.

[18] Song, Y., Swedlund, P. J. & Singhal, N. (2008). *Environ. Sci. Technol.*, *42*, 4008-4013.

[19] Frimmel, F. & Huber, L. (1996). *Environ. Int.*, *22*, 507-517.

[20] Song, Y., Swedlund, P. J., Singhal, N. & Swift, S. (2009). *Environ. Sci. Technol.*, *43*, 7430-7436.

[21] Davis, A. & Bhatnagar, V. (1995). *Chemosphere*, *30*, 243-256.

[22] Arias, M., Barral, M. T. & Mejuto, J. C. (2002). *Chemosphere*, *48*, 1081-1088.

[23] Bargar, J. R., Persson, P. & Brown, G. E. (1999). *Geochim. Cosmochim. Acta*, *63*, 2957-2969.

[24] Noren, K. & Persson, P. J. (2010). *J. Phys. Chem. C*, *114*, 16547-16555.

[25] Lee, S., Nagy, K. L., Park, C. & Fenter, P. (2011). *Environ. Sci. Technol.*, *45*, 9574-9581.

[26] Stietiya, M. H., Wang, J. J. & Roy, A. (2011). *Soil Sci.*, *176*, 464-471.

[27] Dong, D., Hua, X., Li, Y. & Kang, C. (2010). *Biofilms in Natural Aquatic Environment and the Adsorption of Heavy Metals to Them*; Science Press: Beijing.

[28] Lion, L. W., Shuler, M. L. & Ghiorse, W. C. (1988). *CRC Crit. Rev. Environ. Cont.*, *17*, 273-306.

[29] van Hullebusch, E. D., Zandvoort, M. H. & Lens, P. N. L. (2003). *Rev. Environ. Sci. Biotechnol.*, *2*, 9-33.

[30] Beck, A. J., Janssen, F., Polerecky, L., Herlory, O. & de Beer, D. (2009). *Environ. Sci. Technol.*, *43*, 7245-7251.

[31] Pokrovsky, O. S., Feurtet-Mazel, A., Martinez, R. E., Morin, S., Baudrimont, M., Duong, T. & Coste, M. (2010). *Appl. Geochem.*, *25*, 418-427.

[32] Hua, X., Li, M., Su, Y., Dong, D., Guo, Z. & Liang, D. (2012). *J. Hazard. Mater.*, 229-230, 450-454.

[33] Ancion, P. Y., Lear, G. & Lewis, G. D. *Environ. Pollut. 158*, 2738-2745.

[34] Nelson, Y. M., Lion, L. W., Shuler, M. L. & Ghiorse, W. C. (1999). *Limnol. Oceanogr.*, *44*, 1715-1729.

[35] Dong, D., Nelson, Y. M., Lion, L. W., Shuler, M. L. & Ghiorse, W. C. (2000). *Water Res.*, *34*, 427-436.

[36] Dong, D., Hua, X., Li, Y., Zhang, J. & Yan, D. (2003). *Environ. Sci. Technol.*, *37*, 4106-4112.

[37] Dong, D., Liu, L., Hua, X. & Lu, Y. (2007). *Microchem. J.*, *85*, 270-275.

[38] Hua, X., Dong, D., Ding, X., Yang, F., Jiang, X. & Guo, Z. (2012). *Environ. Sci. Pollut. Res.*, *20*, 413-420.

[39] Hua, X., Dong, D., Liu, L., Gao, M. & Liang, D. (2012). *Appl. Geochem.*, *27*, 1005-1012.

[40] Güçlü, K. & Apak, R. J. (2003). *Colloid Interface Sci.*, *260*, 280-290.

[41] Malandrino, M., Abollino, O., Giacomino, A., Aceto, M. & Mentasti, E. J. (2006). *Colloid Interface Sci.*, *299*, 537-546.

[42] Nowack, B. & Sigg, L. J. (1996). *Colloid Interface Sci.*, *177*, 106-121.

[43] Jiann, K., Santschi, P. H. & Presley, B. J. (2013). *Aquat. Geochem.*, *19*, 173-193.

[44] Papassiopi, N., Tambouris, S. & Kontopoulos, A. (1999). *Water Air Soil Pollut.*, *109*, 1-15.

[45] Lestan, D., Luo, C. & Li, X. (2008). *Environ. Pollut.*, *153*, 3-13.

[46] Xue, H., Sigg, L. & Kari, F. G. (1995). *Environ. Sci. Technol.*, *29*, 59-68.

[47] Wang, X. & Feng, J. (2007). *Environ. Manage.*, *40*, 314-322.

[48] Gray, B. R., Hill, W. R. & Stewart, A. J. (2001). *Environ. Pollut.*, *112*, 61-71.

[49] Schäfer, J., Norra, S., Klein, D. & Blanc, G. J. (2009). *Soils Sediments*, *9*, 374-392.

[50] Dong, D., Ji, L., Hua, X., Li, Y. & Zheng, N. (2004). *Chem. J. Chin. Univ.*, *25*, 247-251.

[51] Serre, S. D. & Silcox, G. D. (2000). *Ind. Eng. Chem. Res.*, *39*, 1723-1730.

[52] Mahmoodi, N. M. (2013). *Water Air Soil Pollut.*, *224*, 1419.

[53] Nowack, B. & Sigg, L. (1995). *Environ. Sci. Technol.*, *29*, 3070-3071.

[54] Nowack, B., Kari, F. G. & Krüger, H. G. (2001). *Water Air Soil Pollut.*, *125*, 243-257.

[55] Komárek, M., Tlustoš, P., Száková, J., Chrastný, V. & Balík, J. (2007). *Plant Soil Environ.*, *53*, 216-224.

[56] Tsang, D. C. W., Zhang, W. & Lo, I. M. C. (2007). *Chemosphere*, *68*, 234-243.

[57] Li, X. D. & Schwartz, F. W. (2004). *J. Contam. Hydrol.*, *68*, 39-53.

[58] Tebo, B. M., Bargar, J. R., Clement, B. G., Dick, G. J., Murray, K. J., Parker, D., Verity, R. & Webb, S. M. (2004). *Annu. Rev. Earth Planet. Sci.*, *32*, 287-328.

[59] Duckworth, O. W. & Sposito, G. (2007). *Chem. Geol.*, *242*, 497-508.

[60] Speight, J. G. (2005). *Lange's Handbook of Chemistry* (16th ed.); McGraw-Hill: New York.

In: EDTA
Editor: Andris Molnar

ISBN: 978-1-62808-146-6

Chapter 6

MANAGEMENT OF EDTA-CONTAINING AQUEOUS EFFLUENT: ENVIRONMENTAL CONCERNS AND REMEDIES

Zinnat A. Begum,[1] Ismail M. M. Rahman[1,2,*] and Hiroshi Hasegawa[3,†]

[1]Graduate School of Natural Science and Technology, Kanazawa University, Kakuma, Kanazawa, Japan
[2]Department of Applied and Environmental Chemistry, University of Chittagong, Chittagong, Bangladesh
[3]Institute of Science and Engineering, Kanazawa University, Kakuma, Kanazawa, Japan

ABSTRACT

Ethylenediaminetetraacetic acid, commonly abbreviated as EDTA, has received widespread acclaim among the aminopolycarboxylate chelants for its superior metal binding capability to generate soluble metal-EDTA complexes having diminished reactivity. EDTA has been and continues to be used in a variety of industrial applications, or otherwise. The objective is either to mask the hindering effect of unwanted metal-species or the separation of the toxic metals. Management of EDTA-containing aqueous effluents has received considerable attention because it not only possesses

* E-mail: I.M.M.Rahman@gmail.com
† E-mail: hhiroshi@t.kanazawa-u.ac.jp; Tel/Fax: +81-76-234-4792.

the potential to affect metal bioavailability but also can remobilize the metals from sediments and aquifers. Besides, the persistent nature of EDTA in the aqueous environment due to the slower rate of biodegradability is the largest issue of concern considering the consequent risk to the potable water resources. The proposed remedies are primarily focused on the destruction of EDTA in the aqueous effluent before being discharged to the natural surroundings. The recovery of EDTA from the effluent for subsequent application is also proposed as a remedial option to minimize the probable risk from EDTA. Within the scope of this chapter, we will provide a brief overview, in light of the recent findings, about the possible environmental concerns related to the discharge of EDTA-containing effluent and the proposed options to compromise the risks.

BACKGROUND

Ethylenediaminetetraacetic acid and its salts are commonly referred to as EDTA and is a well-known chelating agent (or chelant) for various applications since its commercial production in 1939 [1]. Being a hexadentate chelant, EDTA is capable of forming complexes stoichiometrically with virtually every metal in the periodic table [2]. The parent EDTA molecule is a substituted diamine, and usually marketed worldwide as its sodium salts under 30 different trademarks [3]. The usage of EDTA by industry as percentages of the world market can be mentioned from Oviedo and Rodríguez [3]: detergents, 33%; water treatment, 18%; pulp and paper industry, 13%; photography, 5%; metal cleaning, 5%; cosmetics, foodstuffs and pharmaceuticals, 5%; agrochemicals, 4%; textile industry, 4%; printing inks, 3%; concrete admixtures, 2%; and miscellaneous, 12%. An overall scenario shows that the global consumption of EDTA is huge and increasing; *e.g.*, the annual EDTA consumption in Europe was 26,000 tons in 1992 [4], the value increased to 32,550 tons in 1997 [5], and it was 34,550 tons in 1999 [1]. A summary of the uses of EDTA is listed in Table 1.

INTERACTION OF EDTA AND METAL SPECIES IN THE AQUEOUS MATRIX

The pathways of a pollutant from the manufacturing plants to the environment and into humans follow manifold routes.

Table 1. Summary of the EDTA applications

Applications	Purpose	Reference
Detergents	Inhibition of water hardness and scale formation	[7, 8]
Photo industry	Oxidizing agent	[9]
Pulp and paper industry	Masking of aqueous metals in the bleaching process	[10, 11]
Metal finishing	Electroless metal plating	[12, 13]
Textile industry	Bleach stabilizer	[11, 14]
Soil remediation	Extraction of heavy metals, and to enhance phytoremediation	[15-18]
Waste management	Nuclear reactor decontamination and waste processing	[19]
Agriculture	Transporter of nutrient metals (*e.g.*, Fe, Zn, and Cu)	[20, 21]
Healthcare	Treatment of metal poisoning, heart disease and high blood pressure and in dental procedures	[22-25]
Consumer applications	Stabilizer; Prevent catalytic reactions	[26, 27]

However, the predominant environmental release of EDTA occurs via the wastewaters attributable to the typical application trend of EDTA as the aqueous solution [3]. The toxicities and environmental behavior of any chelant and its metal complexes in the aqueous matrix depends on the relative species distributions, which can be predicted from the metal-chelant equilibria using the stability constant values. The speciation calculation of any metal-chelant system helps to estimate the dominating chemical forms of metal in the solution, and the pH range for stable complex formation. Moreover, the masking ability of a chelant can also be predicted as the log[M] value of the free metal ion at any pH is available from such speciation calculations [6].

The acid dissociation constants (pK_a) of EDTA (matrix: H_2O; ionic strength, I = 0.1 mol·dm^{-3}; T = 25 °C) were reported in the NIST database [28], which are as follows: 2.00, 2.69, 6.13, 10.37. The percentage distribution of the different protonation stages of EDTA at the aforementioned conditions is shown in Figure 1a, which illustrates the occurrence of the first and second protonation of L^{4-} to H_2L^{2-} at the amino nitrogen atoms. The HL^{3-} remains as the dominant species at pH 7.0–9.5 (88–97%) for EDTA. In the range of acidic pH, the next protonation's of EDTA (H_3L^- to H_4L) occur at the oxygen atoms of the carboxylate groups.

The pH dependent species distribution curves drawn in the range of pH 2 to 11 (ionic strength, $I = 0.1$ mol·dm^{-3}) for EDTA complexation with different metal ions (Cd^{2+}, Cu^{2+}, Pb^{2+}, Hg^{2+}, Ni^{2+}, Zn^{2+}, and Fe^{3+}) are shown in Figure 1b to 1h. The United States Environmental Protection Agency mentioned the Cd^{2+}, Cu^{2+}, Pb^{2+}, Hg^{2+}, Ni^{2+}, Zn^{2+} metal species as the priority pollutant to be controlled in the effluent before being released to the water bodies [29]. The Fe(III)EDTA is an important EDTA species released from wastewater treatment plants and comprises 20 to 90% of total EDTA in the effluents [30, 31]. The stability constants of the EDTA complexes with those metals are listed in Table 2.

The complexation among the metal (M) and EDTA (L) species were observed to start from the initial pH ~2 with the formation of MH_2L species in the M(II)–EDTA systems, except that of Ni(II)–EDTA, while it is MH_2L^+ for Fe(III)–EDTA systems. The first species MH_2L or MH_2L^+ was expected to form by the coordination of weakly basic imine nitrogen atom and the deprotonated carboxylic group in EDTA, where the other remaining coordination sites of the metal ions may be occupied by the water molecules. As the pH increases, coordinated water molecules are replaced by the deprotonated carboxylic groups in EDTA which leads to the formation of the MHL^-, ML^{2-}, $M(OH)L^{3-}$ complexes (M = Cd^{2+}, Cu^{2+}, Pb^{2+}, Hg^{2+}, Ni^{2+}, Zn^{2+}) or MHL, ML^-, $M(OH)L^{2-}$ (M = Fe^{3+}) (L = EDTA). A list of the dominant species formed within the ML systems and the corresponding pH ranges are provided in Table 3. Above 84% formation of all the dominant species were occurred virtually in whole pH range, while the ML^{2-} (M = Cd^{2+}, Cu^{2+}, Pb^{2+}, Hg^{2+}, Ni^{2+}, Zn^{2+}) or ML^- (M = Fe^{3+}) (L = EDTA) remain the common dominant species.

The other competing factors *e.g.* pH or interfering effects from other species, are ignored while the stability constant values for ML systems are derived. Therefore, the term 'conditional stability constant' ($\log K'_{ML}$) is defined considering the effects of side reactions that may occur during the complexation of chelant with metal ions such as, the effects of chelant protonation and of hydrolysis that prevails when a metal ion is in solution [33].

The $\log K'_{ML}$ values of the ML complexes (M = Cd^{2+}, Cu^{2+}, Pb^{2+}, Hg^{2+}, Ni^{2+}, Zn^{2+}, or Fe^{3+}; L = EDTA) in the aqueous medium at the ionic strength, $I = 0.1$ mol·dm^{-3} and at 25 °C as a function of pH (2–12) were calculated using HySS2009 program [32], and given in Table 2.

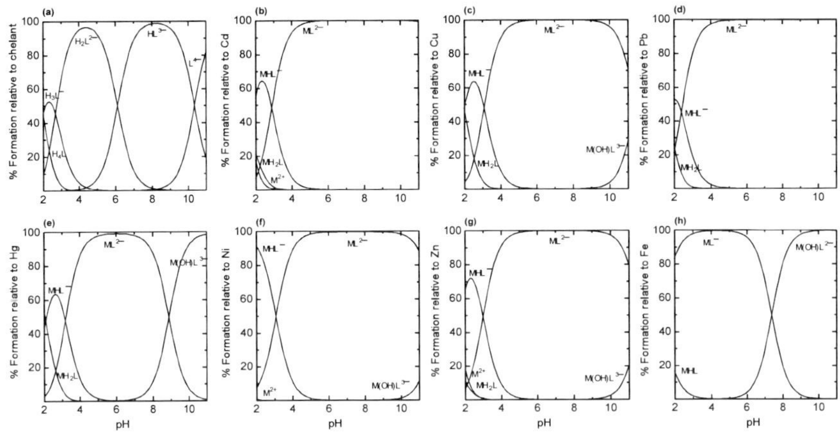

Figure 1. The percentage distribution of different protonation stages and the species distribution curves for ML systems (M = Cd^{2+}, Cu^{2+}, Pb^{2+}, Hg^{2+}, Ni^{2+}, Zn^{2+}, or Fe^{3+}; L = EDTA) in aqueous medium at ionic strength, $I = 0.1$ mol·dm^{-3} and at 25 °C.

In the natural aqueous systems, the predicted equilibrium conditions are rather complicated due to the presence of other competing natural or anthropogenic chelants, ionic species, suspended particles and sediments. An obvious alternation in the equilibrium is expected if another metal-chelant complex or metal ion is added in the solution. The interference in the kinetics of the coordination reaction is a common occurrence during the complexation between EDTA and metal ions. A strong decrease in the EDTA complexation rate might also occur because of the binding with Ca and Mg in hard waters [30].

Concerns Intended to EDTA Release in the Environment

Degradability

EDTA, being a multidentate chelant, is non-selective in nature and the elements other than the target one simultaneously compete for the binding sites. An excess amount of chelant is, hence, added to minimize such competing effect and ensure the sufficient separation of the target metal-species.

Table 2. Stability constants ($\log K_{ML}$) and conditional stability constants ($\log K'_{ML}$) of the ML complexes ($M = Cd^{2+}$, Cu^{2+}, Pb^{2+}, Hg^{2+}, Ni^{2+}, Zn^{2+}, or Fe^{3+}; L = EDTA) [a]

Metal	$\log K_{ML}$	$\log K'_{ML}$ at different pH [b]										
		2	3	4	5	6	7	8	9	10	11	12
Cd(II)	16.5	4.1	6.1	8.0	10.0	11.8	13.1	14.1	15.0	15.5	14.4	12.0
Cu(II)	18.78	6.7	8.5	10.3	12.3	14.0	15.3	16.1	16.2	15.3	13.4	10.5
Pb(II)	18	7.5	10.0	11.6	12.6	13.6	14.6	15.2	15.3	14.6	12.8	10.0
Hg (II)	21.5	9.5	11.0	11.4	11.3	11.1	10.5	9.6	8.8	8.5	7.9	7.0
Ni(II)	18.4	6.0	8.1	9.9	11.9	13.7	15.0	16.0	16.9	16.8	15.1	12.7
Zn(II)	16.5	4.1	6.1	8.0	10.0	11.8	13.1	14.1	14.1	13.0	11.1	8.8
Fe(III)	25.1	11.5	13.8	14.6	14.7	14.4	13.2	11.9	10.7	9.0	6.5	3.6

[a] Matrix: H_2O; ionic strength, $I = 0.1$ mol·dm^{-3}; $T = 25$ °C [28].

[b] The change in conditional stability constants of the ML complexes in terms of the solution pH is calculated with the aid of the computer program HySS2009 [32].

Table 3. The dominant species in the ML systems (M = Cd^{2+}, Cu^{2+}, Pb^{2+}, Hg^{2+}, Ni^{2+}, Zn^{2+}, or Fe^{3+}; L = EDTA) [a]

Metal Species	Dominant Species	pH range	% formation
Cd(II)	$CdEDTA^{2-}$	4.0~	> 91
Cu(II)	$CuEDTA^{2-}$	4.0–10.4	> 91
Pb(II)	$PbEDTA^{2-}$	4.0~	> 97
Hg (II)	$HgEDTA^{2-}$	4.0–8.0	> 88
Ni(II)	$NiEDTA^{2-}$	4.0–11.0	> 89
Zn(II)	$ZnEDTA^{2-}$	4.0–10.5	> 91
Fe(III)	$FeEDTA^{-}$	2.0–6.6	> 84

[a] Matrix: H_2O; ionic strength, I = 0.1 mol·dm^{-3}; T = 25 °C. The dominant species information is available from the pH dependent species distribution curves for the ML systems that were drawn within the range of pH 2 to 11 (Figure 1) with the aid of the computer program HySS2009 [32].

The anthropogenic input of EDTA in the natural aqueous systems has been detected from few to 100 μg L^{-1} [34].

The rate of degradation of a pollutant, when enter the environment, by both abiotic and abiotic processes is a key issue to estimate its fate. The EDTA is not degraded by conventional biological and physicochemical methods and long considered non-biodegradable. In recent years, however, it has been demonstrated that some specially enriched bacterial cultures possess the ability to oxidize EDTA [5]. The photo- and chemical degradation process also contributes to some extent in the long-term decomposition behavior of EDTA and its complexes [31, 35]. In summary, the rate of photo–, chemo– and biodegradability of EDTA is poor and, hence, the exposure effects are likely to persist for a long-time [30, 36-39].

Ecotoxic Effects

EDTA has the potential to contribute to eutrophication by increasing the total nitrogen content when mineralized in the natural water bodies. EDTA also able to increase the availability of the inorganic phosphates via desorption from its Fe-bound forms as present in sediments, and thereby influences the algal growth in interstitial waters [40-42].

When released into aquatic environments, EDTA may induce the remobilization of metal ions from soils and sediments into the water phase [1,

36], thus extend the residence time of the metals. The factors responsible for the EDTA-induced metal remobilization are metal and EDTA concentrations, pH, and the presence of other co-existing complexing agents and cations, whether anthropogenic or natural [43]. Additionally, in most cases, the toxicity threshold values of EDTA increase with metal complexation [34, 44, 45].

Guideline Value for EDTA in Drinking-Water

The toxicity of EDTA and its calcium/sodium salts have been evaluated based on the available toxicological studies, and a guideline value for EDTA in drinking-water is derived by World Health Organization [46]. The value for EDTA (free acid) is 600 μg L^{-1} assuming a 60-kg adult ingesting 2 L of drinking-water per day.

Pre-Release Management of EDTA-Containing Effluent Water

The legislative agencies have become concerned about eco-environmental consequences due to the increasing use of EDTA, and increasingly stringent environmental regulations have been imposed [47, 48]. Therefore, treatment of EDTA-containing wastewaters is a prerequisite before they can be safely discharged depending on the characteristics and concentrations of both the added chelant and metals in the source [49].

The degradation treatment of the EDTA-containing effluent water is considered when the concentration falls below 1 mmol L^{-1} [50], and several methods have been proposed either to eradicate or at least for minimizing the concentration of EDTA in discharge waters [10, 51-55]. The recovery and reuse of EDTA and metals become the main concern for concentrations above 5 mmol L^{-1} in solution [56] because the cost of chelants is a critical issue surrounding their use in metal ion sequestration [18, 57, 58]. Some potential techniques proposed for the purpose are as follows: (a) an electrochemical reduction treatment followed by membrane separation [56, 59]; (b) a precipitation treatment with zero-valent metals [60]; (c) the addition of suitable reagents (e.g., NaOH, $Ca(OH)_2$, Na_2S, $FeSO_4$, $FeCl_3$, NaH_2PO_4, Na_2HPO_4, or diethyldithiocarbamate) [57, 58, 61-66]; (d) the application of a macrocycle-immobilized solid-phase extraction system, commonly known as molecular recognition technology gel [67, 68]. The operational problems, *e.g.* membrane fouling, membrane degradation, extensive costs are some drawbacks of the proposed treatment techniques [57, 58, 65]. Furthermore, most of those techniques are based on equimolar concentrations of metals and EDTA in solution [56, 57, 62], while effluent waters are often characterized by a large excess of free EDTA in mixtures [18, 65].

CONCLUSION

The excellent metal-binding capacities of EDTA facilitate its extensive usage in various metal-masking processes in industries, in households, or in agriculture. The pre- and post-toxicity of EDTA after its ultimate release to the aquatic environment and its high persistence nature evoke concerns, and have been pointed out in this article. The predicted equilibria between the EDTA and the metal species in the aqueous matrix is explained. Several treatment techniques with a principal focus on the degradation of EDTA at the pre-release step have been proposed, which have been discussed including their limitations. The conclusive fact is the requirement of an extensive research effort to develop an economically beneficial technique that ensures the recycling of processed water, including EDTA.

ACKNOWLEDGMENT

This research was partially supported by the Grants-in-Aid for Scientific Research (24310056 and 24·02029) from the Japan Society for the Promotion of Science.

REFERENCES

[1] Nowack, B.; VanBriesen, J. M., Chelating agents in the environment. In *Biogeochemistry of Chelating Agents,* Nowack, B.; VanBriesen, J. M., Eds. American Chemical Society: Washington, DC, 2005; pp 1–18.

[2] Chaberek, S.; Martell, A. E., *Organic sequestering agents; a discussion of the chemical behavior and applications of metal chelate compounds in aqueous systems.* Wiley: New York, 1959.

[3] Oviedo, C.; Rodríguez, J., EDTA: the chelating agent under environmental scrutiny. *Quim. Nova,* 2003, *26*, 901–905.

[4] Henneken, L.; Nörtemann, B.; Hempel, D. C., Influence of physiological conditions on EDTA degradation. *Appl. Microbiol. Biotechnol.,* 1995, *44*, 190–197.

[5] Nortemann, B., Biodegradation of EDTA. *Appl. Microbiol. Biotechnol.,* 1999, *51*, 751-759.

[6] Popov, K. I.; Wanner, H., Stability constants data sources: Critical evaluation and application for environmental speciation. In *Biogeochemistry of Chelating Agents,* Nowack, B.; VanBriesen, J. M., Eds. American Chemical Society: Washington, DC, 2005; pp 50–73.

[7] Steber, J.; Wiebel, F., Laundry detergents. In *Ullmann's Encyclopedia of Industrial Chemistry,* Wiley-VCH Verlag: Weinheim, 2000.

[8] Nitsch, C.; Heitland, H.-J.; Marsen, H.; Schlüssler, H.-J., Cleansing agents. In *Ullmann's Encyclopedia of Industrial Chemistry,* Wiley-VCH Verlag: Weinheim, 2000.

[9] Keller, K.; Kampfer, H.; Matejec, R.; Lapp, O.; Krafft, W.; Frenken, H.; Lührig, H.; Scheerer, R.; Heilmann, M.; Meckl, H.; Bergthaller, P.; Hübner, D.; Wolff, E.; Morcher, B.; Zahn, W.; Buschmann, H.; Blank, R.; Tromnau, R.; Plamper, J.; Seiler, A.; Nieswandt, K.; Boie, I.; Moisar, E.; Winiker, R.; Schellenberg, M.; Ketellapper, L., Photography. In *Ullmann's Encyclopedia of Industrial Chemistry,* Wiley-VCH Verlag: Weinheim, 2000.

[10] Pirkanniemi, K.; Metsärinne, S.; Sillanpää, M., Degradation of EDTA and novel complexing agents in pulp and paper mill process and waste waters by Fenton's reagent. *J. Hazard. Mater.,* 2007, *147*, 556–561.

[11] Jorgensen, R. J.; Nyengaard, N. R.; Daniel, R. C. W.; Mellau, L. S. B.; Enemark, J. M. D., Induced hypocalcaemia by Na(2)EDTA infusion. A review. *J. Vet. Med. A,* 1999, *46*, 389–407.

[12] Lin, Y.-M.; Yen, S.-C., Effects of additives and chelating agents on electroless copper plating. *Appl. Surf. Sci.,* 2001, *178*, 116–126.

[13] Mallory, G. O.; Hajdu, J. B., *Electroless Plating Fundamentals and Applications.* William Andrew Publishing: Norwich, NY, 2005.

[14] Leube, H.; Rüttiger, W.; Kühnel, G.; Wolff, J.; Ruppert, G.; Schmitt, M.; Heid, C.; Hückel, M.; Flath, H.-J.; Beckmann, W.; Brossmann, R.; Söll, M.; Sewekow, U., Textile dyeing. In *Ullmann's Encyclopedia of Industrial Chemistry,* Wiley-VCH Verlag: Weinheim, 2000.

[15] Peters, R. W., Chelant extraction of heavy metals from contaminated soils. *J. Hazard. Mater.,* 1999, *66*, 151–210.

[16] Huang, J. W.; Chen, J.; Berti, W. R.; Cunningham, S. D., Phytoremediation of lead-contaminated soils: role of synthetic chelates in lead phytoextraction. *Environ. Sci. Technol.*, 1997, *31*, 800–805.

[17] Doumett, S.; Lamperi, L.; Checchini, L.; Azzarello, E.; Mugnai, S.; Mancuso, S.; Petruzzelli, G.; Del Bubba, M., Heavy metal distribution between contaminated soil and *Paulownia tomentosa*, in a pilot-scale assisted phytoremediation study: influence of different complexing agents. *Chemosphere,* 2008, *72*, 1481–1490.

[18] Leštan, D.; Luo, C. L.; Li, X. D., The use of chelating agents in the remediation of metal-contaminated soils: A review. *Environ. Pollut.*, 2008, *153*, 3–13.

[19] Ayres, J. A., Decontamination of Nuclear Reactors and Equipment. Ronald Press Co.: New York, 1970.

[20] Abdulla, I.; Smith, M. S., Influence of chelating agents on the concentration of some nutrients for plants growing in soil under acid and under alkaline conditions. *J. Sci. Food Agric.*, 1963, *14*, 98–109.

[21] Stewart, I., Chelation in the absorption and translocation of mineral elements. *Ann. Rev. Plant Physio.*, 1963, *14*, 295–310.

[22] Mayfield, L.; Söderholm, G.; Norderyd, O.; Attström, R., Root conditioning using EDTA gel as an adjunct to surgical therapy for the treatment of intraosseous periodontal defects. *J. Clin. Periodontol.*, 1998, *25*, 707–714.

[23] Olszewer, E.; Carter, J. P., EDTA chelation therapy in chronic degenerative disease. *Med. Hypotheses,* 1988, *27*, 41–49.

[24] Chappell, L. T.; Janson, M., EDTA chelation therapy in the treatment of vascular disease. *J. Cardiovasc. Nurs.*, 1996, *10*, 78–86.

[25] Sloth-Nielsen, J.; Guldager, B.; Mouritzen, C.; Lund, E. B.; Egeblad, M.; Nørregaard; O.; Jørgensen, S. J.; Jelnes, R., Arteriographic findings in EDTA chelation therapy on peripheral arteriosclerosis. *Am. J. Surg.*, 1991, *162*, 122–125.

[26] Whittaker, P.; Vanderveen, J. E.; Dinovi, M. J.; Kuznesof, P. M.; Dunkel, V. C., Toxicological profile, current use, and regulatory issues on EDTA compounds for assessing use of sodium iron EDTA for food fortification. *Regul. Toxicol. Pharmacol.*, 1993, *18*, 419–427.

[27] Kemmei, T.; Kodama, S.; Yamamoto, A.; Inoue, Y.; Hayakawa, K., Determination of sequestering agents in cosmetics and synthetic detergents by high-performance liquid chromatography with ultraviolet detection. *J. Chromatogr., A* 2007, *1171*, 63–68.

[28] Martell, A. E.; Smith, R. M.; Motekaitis, R. J., NIST Standard Reference Database 46: NIST Critically Selected Stability Constants of Metal Complexes Database (Version 8.0 For Windows). In Texas AandM University: College Station, TX, 2004.

[29] U.S. EPA The Clean Water Act (Priority Pollutants: Appendix A to 40 CFR Part 423); U.S. Environmental Protection Agency: Washington, D.C, 2012.

[30] Nowack, B., Environmental chemistry of aminopolycarboxylate chelating agents. *Environ. Sci. Technol.,* 2002, *36*, 4009–4016.

[31] Kari, F. G.; Giger, W., Speciation and fate of ethylene-diaminetetraacetate (EDTA) in municipal wastewater treatment. *Water Res.,* 1996, *30*, 122–134.

[32] Alderighi, L.; Gans, P.; Ienco, A.; Peters, D.; Sabatini, A.; Vacca, A., Hyperquad simulation and speciation (HySS): A utility program for the investigation of equilibria involving soluble and partially soluble species. *Coordin. Chem. Rev.,* 1999, *184*, 311–318.

[33] Davidge, J.; Thomas, C. P.; Williams, D. R., Conditional formation constants or chemical speciation data? *Chem. Speciation Bioavailability,* 2001, *13*, 129–134.

[34] Sillanpää, M. E. T., Distribution and fate of chelating agents in the environment. In *Biogeochemistry of Chelating Agents, Nowack, B.; VanBriesen, J.* M., Eds. American Chemical Society: Washington, DC, 2005; pp 226–233.

[35] Frank, R.; Rau, H., Photochemical transformation in aqueous solution and possible environmental fate of ethylenediaminetetraacetic acid (EDTA). *Ecotoxicol. Environ. Saf.,* 1990, *19*, 55–63.

[36] Means, J. L.; Kucak, T.; Crerar, D. A., Relative degradation rates of NTA, EDTA and DTPA and environmental implications. *Environ. Pollut. B* 1980, *1*, 45–60.

[37] Egli, T., Biodegradation of metal-complexing aminopolycarboxylic acids. *J. Biosci. Bioeng.,* 2001, *92*, 89–97.

[38] Kari, F. G.; Giger, W., Modeling the photochemical degradation of ethylenediaminetetraacetate in the river Glatt. *Environ. Sci. Technol.,* 1995, *29*, 2814–2827.

[39] Nörtemann, B., Biodegradation of chelating agents: EDTA, DTPA, PDTA, NTA, and EDDS. In *Biogeochemistry of Chelating Agents,* Nowack, B.; VanBriesen, J. M., Eds. American Chemical Society: Washington, DC, 2005; pp 150–170.

[40] Erel, Y.; Morgan, J. J., The relationships between rock-derived lead and iron in natural waters. *Geochim. Cosmochim. Ac.,* 1992, *56*, 4157–4167.

[41] Hering, J. G.; Morel, F. M. M., Kinetics of trace metal complexation: Role of alkaline-earth metals. *Environ. Sci. Technol.,* 2002, *22*, 1469–1478.

[42] Horstmann, U.; Gelpke, N., Algal growth stimulation by chelatisation risks associated with complexants in P-free washing agents. *Rev. Intl. Oceanogr. Med.,* 1991, *260*, 101–104.

[43] Sillanpää, M., Environmental fate of EDTA and DTPA. *Rev. Environ. Contam. Toxicol.,* 1997, *152*, 85–111.

[44] Sillanpää, M.; Oikari, A., Assessing the impact of complexation by EDTA and DTPA on heavy metal toxicity using microtox bioassay. *Chemosphere,* 1996, *32*, 1485–1497.

[45] Sorvari, J.; Sillanpää, M., Influence of metal complex formation on heavy metal and free EDTA and DTPA acute toxicity determined by *Daphnia magna. Chemosphere,* 1996, *33*, 1119–1127.

[46] WHO Edetic acid (EDTA) in Drinking-water (In: *Guidelines for drinking-water quality;* 2nd ed. Addendum to Vol. 2. Health criteria and other supporting information); WHO/SDE/WSH/03.04/58; World Health Organization (WHO): Geneva, 1998.

[47] van Ginkel, C. G.; Geerts, R., Full-Scale biological treatment of industrial effluents containing EDTA. In *Biogeochemistry of Chelating Agents*, Nowack, B.; VanBriesen, J. M., Eds. American Chemical Society: Washington, DC, 2005; pp 195–203.

[48] Grundler, O. J.; van der Steen, A. T. M.; Wilmot, J., Overview of the European risk assessment on EDTA. In *Biogeochemistry of Chelating Agents,* Nowack, B.; VanBriesen, J. M., Eds. American Chemical Society: Washington, DC, 2005; pp 336–347.

[49] Juang, R. S.; Wang, S. W.; Lin, L. C., Simultaneous recovery of EDTA and lead(II) from their chelated solutions using a cation exchange membrane. *J. Membr. Sci.,* 1999, *160*, 225–233.

[50] Juang, R. S.; Wang, S. W., Electrolytic recovery of binary metals and EDTA from strong complexed solutions. *Water Res.,* 2000, *34*, 3179–3185.

[51] Bucheli-Witschel, M.; Egli, T., Environmental fate and microbial degradation of aminopolycarboxylic acids. *FEMS Microbiol., Rev.* 2001, *25*, 69–106.

[52] Krapfenbauer, K.; Getoff, N., Comparative studies of photo- and radiation-induced degradation of aqueous EDTA. Synergistic effects of

oxygen, ozone and TiO_2 (acronym: CoPhoRaDe/EDTA). *Radiat. Phys. Chem.,* 1999, *55*, 385–393.

[53] Rämö, J.; Sillanpää, M., Degradation of EDTA by hydrogen peroxide in alkaline conditions. *J. Clean. Prod.,* 2001, *9*, 191–195.

[54] Sillanpää, M.; Pirkanniemi, K., Recent developments in chelate degradation. *Environ. Technol.,* 2001, *22*, 791–801.

[55] Muñoz, F.; von Sonntag, C., The reactions of ozone with tertiary amines including the complexing agents nitrilotriacetic acid (NTA) and ethylenediaminetetraacetic acid (EDTA) in aqueous solution. *J. Chem. Soc. Perk. T,. 2* 2000, 2029–2033.

[56] Juang, R. S.; Wang, S. W., Metal recovery and EDTA recycling from simulated washing effluents of metal-contaminated soils. *Water Res.,* 2000, *34*, 3795–3803.

[57] Kim, C.; Ong, S.-K., Recycling of lead-contaminated EDTA wastewater. *J. Hazard. Mater.,* 1999, *69*, 273–286.

[58] Lim, T. T.; Chui, P. C.; Goh, K. H., Process evaluation for optimization of EDTA use and recovery for heavy metal removal from a contaminated soil. *Chemosphere,* 2005, *58*, 1031–1040.

[59] Arévalo, E. F.; Stichnothe, H.; Thöming, J.; Calmano, W., Evaluation of a leaching process coupled with regeneration/recycling of the extractant for treatment of heavy metal contaminated solids. *Environ. Technol.,* 2002, *23*, 571–581.

[60] Lee, C. C.; Marshall, W. D., Recycling of complexometric extractants to remediate a soil contaminated with heavy metals. *J. Environ. Monitor.,* 2002, *4*, 325–329.

[61] Tünay, O.; Kabdasli, N. I., Hydroxide precipitation of complexed metals. *Water Res.,* 1994, *28*, 2117–2124.

[62] Chang, L.-Y., A waste minimization study of a chelated copper complex in wastewater – Treatability and process analysis. *Waste Manag.,* 1995, *15*, 209–220.

[63] Steele, M. C.; Pichtel, J., Ex-situ remediation of a metal-contaminated superfund soil using selective extractants. *J. Environ. Eng.-ASCE,* 1998, *124*, 639–645.

[64] Xie, T.; Marshall, W. D., Approaches to soil remediation by complexometric extraction of metal contaminants with regeneration of reagents. *J. Environ. Monitor.,* 2001, *3*, 411–416.

[65] Di Palma, L.; Ferrantelli, P.; Merli, C.; Biancifiori, F., Recovery of EDTA and metal precipitation from soil flushing solutions. *J. Hazard. Mater.,* 2003, *103*, 153–168.

[66] Hong, P. K. A.; Li, C.; Banerji, S. K.; Regmi, T., Extraction, Recovery, and Biostability of EDTA for Remediation of Heavy Metal-Contaminated Soil. *J. Soil Contam.*, 1999, *8*, 81–103.
[67] Hasegawa, H.; Rahman, I. M. M.; Nakano, M.; Begum, Z. A.; Egawa, Y.; Maki, T.; Furusho, Y.; Mizutani, S., Recovery of toxic metal ions from washing effluent containing excess aminopolycarboxylate chelant in solution. *Water Res.*, 2011, *45*, 4844–4854.
[68] Hasegawa, H.; Rahman, I. M. M.; Kinoshita, S.; Maki, T.; Furusho, Y., Non-destructive separation of metal ions from wastewater containing excess aminopolycarboxylate chelant in solution with an ion-selective immobilized macrocyclic material. *Chemosphere*, 2010, *79*, 193–198.

INDEX

A

absorption spectroscopy, 109
acetaldehyde, 107
acetic acid, 18
acetone, 31
acidic, 100, 103, 165
acidity, 6, 7
acrylate, 28
activation energy, 95
adamantane, 19, 23, 40, 41
additives, 11, 37, 172
adhesion, 120, 124, 126, 128, 129, 130, 137, 138
adjunctive therapy, 132
adsorption, vii, ix, 10, 104, 108, 124, 126, 141, 142, 143, 144, 145, 146, 148, 150, 151, 152, 153, 154, 155, 156, 158
adsorption isotherms, 143
agencies, 170
aggregation, 7, 17, 24, 34
agriculture, 171
alanine, 109
alcohols, vii, 1, 3, 6, 103
algorithm, 56
alkane, 42
alkylation, 27
amine(s), vii, 1, 3, 4, 5, 8, 12, 13, 14, 26, 27, 39, 176
amino, 6, 11, 16, 20, 23, 24, 28, 29, 32, 35, 41, 165
amino acid(s), 28, 35
amino groups, 6, 29
aminopolycarboxylate, x, 37, 163, 174, 177
ammonia, 59, 60
ammonium, 8
amplitude, 54
anatase, 97
aniline, 31
annealing, 49
aqueous solutions, 6, 13, 22, 26, 31, 107, 114
aquifers, x, 164
aromatic compounds, 94
aromatic rings, 2
arrest, ix, 119
arteriosclerosis, 173
atmosphere, 49
atoms, 32, 48, 103, 106, 165
attachment, ix, 15, 16, 119, 120, 121, 124, 126, 128, 130, 131, 132, 133, 134, 135, 137, 138
autonomy, 46

B

bacteria, ix, 119, 121, 136
bacterial cells, 146
band gap, 97, 112

Bangladesh, 163
Beijing, 159
benefits, 134
bicarbonate, 109, 113
bile, 16, 17
bile acids, 4, 14, 15, 34, 38, 40
bioassay, 175
bioavailability, x, 164
biochemical processes, 25
biodegradability, x, 143, 164, 169
bioincompatible, ix, 120
biological activity, 18, 144
biological processes, 142
bleaching, 165
blood, 15, 124, 126, 127
bonds, 16, 19, 30, 63
bone, ix, 119, 120, 128, 129, 130, 132, 133, 138
bone form, 132, 133
breakdown, ix, 119, 120
burnout, 63

C

Ca^{2+}, 34, 35, 59, 60
cadmium, vii, x, 107, 109, 110, 112, 113, 114, 141, 143, 144, 145, 146, 147, 148, 149, 150, 151, 152, 153, 154, 155, 156, 157, 158
calcination, viii, 45, 50, 58, 59, 61, 62, 64, 66, 69, 81, 86
calcination temperature, 58, 59, 61, 64, 66, 81
calcium, 121, 126, 170
calculus, 121
calorimetry, 22
candidates, 97
carbon, 4, 12, 16, 17, 20, 95, 108, 144
carbon atoms, 4, 17, 20
carboxyl, 29
carboxylic acid(s), 25, 94
carboxylic groups, 2, 3, 16, 30, 31, 60, 166
catalyst, 96, 97
cation, 32, 175
cell membranes, 25
ceramic(s), 46, 47, 49, 50, 51, 52, 53, 58, 85, 88
ceramic materials, 49, 50
cerium, 41
CFR, 174
chelates, 6, 7, 8, 12, 13, 15, 20, 26, 29, 35, 36, 38, 39, 173
chelating ability, vii, 1, 2, 10, 30
chemical, viii, ix, 38, 45, 50, 58, 60, 61, 73, 88, 92, 94, 97, 101, 102, 103, 104, 119, 121, 122, 131, 133, 138, 139, 142, 143, 165, 169, 171, 174
chemical degradation, 169
chemical properties, 38, 97
chemical reactions, 94
chemical stability, 61, 94
China, 143, 144, 158
chloroform, 31
cholesterol, 26, 39, 42
cholic acid, 14, 15, 17, 39
chromium, 99, 100
circulation, 14
classes, 46, 50
Clean Water Act, 174
cleaning, 143, 164
cleavage, 12, 35
clinical trials, 133, 134
clustering, 106
CO_2, 63, 105, 107
coal, 11, 37
coatings, vii, ix, 141, 142, 143, 144, 145, 146, 147, 148, 149, 150, 151, 152, 153, 154, 155, 156, 158
cobalt, 13, 37, 99
collagen, ix, 119, 121, 122, 123, 124, 131, 132, 134, 137
colonization, 122, 137
color, 31, 146
combined effect, 108
combustion, 59, 144
commercial, 6, 12, 164
communication, 46
communication technologies, 46
comparative analysis, 123
competition, x, 111, 114, 142, 156, 158

complexity, 151
complications, 134
composition, 50, 68, 98, 121, 142
compounds, vii, 2, 3, 5, 6, 8, 9, 10, 14, 15, 16, 17, 18, 19, 27, 29, 30, 31, 32, 34, 37, 63, 109, 171, 173
computer, 168, 169
conditioning, vii, ix, 119, 121, 122, 123, 125, 126, 127, 128, 129, 130, 131, 132, 133, 134, 136, 137, 138, 139, 173
conductivity, 32, 43, 52, 73, 76, 77, 79, 84, 85, 86
configuration, 32, 33
conjugation, 15
connective tissue, ix, 120, 121, 122, 124, 126, 131, 134
consumption, 123, 164
contact time, 108, 109
contaminated soil(s), 143, 172, 173, 176
contamination, 121
control group, 126, 131, 132, 133
controversies, 134
cooperation, 21
coordination, 7, 22, 29, 32, 166, 167
copolymer, 28, 29
copper, 27, 32, 35, 37, 172, 176
correlation, 17, 39, 71, 86, 97, 98, 101, 114, 128, 147
correlation coefficient, 147
cosmetics, 164, 173
cost, 170
Costa Rica, 1
cotton, 6
covering, 123, 128, 130
crystal structure, 33, 35, 49
crystalline, 12, 47, 48, 97
crystallization, 30, 63
crystals, 49
Cuba, 116
culture, 126
cyclodextrins, 2, 19, 20, 26, 42
cysteine, 29
Czech Republic, 91

D

database, 110, 165
debridement, 132, 134
decomposition, viii, 59, 62, 63, 92, 93, 95, 96, 100, 102, 103, 114, 169
decontamination, viii, ix, 91, 92, 93, 98, 101, 119, 165
defects, 130, 131, 132, 133, 134, 139, 173
deficiency, 68
deficit, 69
degradation, vii, viii, 91, 92, 94, 95, 96, 97, 98, 99, 100, 101, 104, 105, 106, 110, 112, 113, 114, 169, 170, 171, 172, 174, 175, 176
degradation mechanism, 96
degradation rate, 95, 174
Degussa, 97, 114
dehiscence, 131, 139
dehydration, 30
dental plaque, ix, 119
dentin, 121, 122, 123, 124, 126, 128, 129, 134, 136, 137, 138
deposition, 129, 138
deposits, 146
depression, 56
depth, 131, 132, 133, 134, 143
derivatives, vii, 1, 2, 3, 5, 6, 8, 9, 11, 12, 13, 16, 17, 18, 19, 20, 22, 23, 25, 26, 27, 28, 31, 33, 34, 35, 36, 37, 38, 39, 40, 41, 43
desorption, 142, 169
destruction, ix, x, 92, 119, 164
detectable, 131
detection, 12, 173
detergents, 6, 36, 164, 172, 173
deviation, 147, 149, 150, 153, 154
diamines, 30, 43
diatoms, 146
dielectric constant, 46, 47, 48, 49, 73, 84
dielectric permittivity, vii, 45, 84
dielectric strength, 73, 86
dielectrics, 46
dimethylformamide, 31
diodes, 84
direct action, 100

direct adsorption, 152
discs, 61
dispersion, 85
dissociation, 6, 102, 114, 156, 165
distribution, 9, 15, 71, 85, 165, 166, 167, 169, 173
divergence, 68, 85
DMF, 27
dogs, 132, 138
DOI, 139
dopants, 49
doping, 49
drug delivery, 25
drug discovery, 39
drugs, 26
drying, 31
dyeing, 172

E

Eastern Europe, 115, 116
effluent(s), vii, x, 163, 164, 166, 170, 175, 176, 177
electric current, 54, 85
electric field, 46, 47, 85
electrical properties, 50
electrode surface, 52
electron, 93, 100, 102, 103, 136
electron microscopy, 136
electrons, viii, 92, 93, 103, 106, 107, 108, 109, 114
emission, 59, 93
enamel, 133, 135, 139
endothermic, 62
endotoxin, ix, 119, 121, 135
endotoxins, 121
energy, 67, 93, 102
energy transfer, 103
engineering, 19, 38
entropy, 23
environment, x, 17, 103, 110, 111, 113, 114, 121, 142, 164, 169, 171, 174
Environmental Protection Agency (EPA), 117, 166, 174
environmental regulations, 170
environments, x, 24, 141, 142, 143, 151, 169
enzyme (s), 7, 34, 36
epithelium, 120, 124, 129
epoxy resins, 26
equilibrium, x, 13, 19, 20, 26, 59, 142, 145, 146, 147, 148, 150, 158, 167
equipment, 98, 101
ester, vii, 2, 5, 6, 8, 10, 13, 14, 25, 32, 35, 36
ester bonds, 2
etching, ix, 120, 121, 122, 123, 124, 128, 129, 131, 132, 136, 137, 138
etching agents, 121, 131
ethanol, 107
ethylene, 3
ethylene glycol, 25, 58, 60, 61
Europe, 164
evidence, 15, 24, 133
excitation, 93, 102
excretion, 15
experimental condition, 15, 126, 153
exposure, ix, 119, 122, 123, 128, 130, 131, 134, 137, 138, 169
extinction, 95
extraction, 10, 13, 36, 38, 122, 128, 130, 170, 172, 176
extracts, 144, 145

F

ferric ion, 34
ferroelectrics, 47
fiber(s), ix, 119, 120, 121, 122, 123, 124, 126, 131, 137, 138
fibrin, 124, 126, 127
fibroblast growth factor, 129, 138
fibroblasts, 126, 129, 138
films, viii, 30, 46, 49
financial support, 34
finite element method, 53
flame, 109, 144
flexibility, 22, 41
food, 173
formula, 4, 64

fouling, 171
fragments, 12, 102
France, 87
freshwater, vii, x, 141
friction, 123, 125
fusion, 49

G

gadolinium, 2
gastrointestinal tract, 15
gel, viii, 29, 45, 50, 58, 60, 61, 62, 63, 64, 66, 67, 68, 69, 71, 72, 73, 74, 75, 77, 78, 79, 80, 81, 82, 83, 85, 86, 122, 123, 124, 127, 128, 129, 130, 132, 134, 136, 137, 138, 139, 171, 173
geometry, 7
Germany, 60
gingival, 120, 126, 131, 132, 135, 138
glucose, 20, 21
glycol, 60
grain boundaries, 50, 51, 52, 53, 56, 72, 73, 75, 77, 81, 82, 84, 85, 86
grain size, 49, 52, 53, 69, 71, 84, 85, 86, 97
graphite, 144
green alga, 146
growth, 6, 42, 46, 67, 121, 128, 129, 130, 135, 138, 169, 175
growth factor, 121, 128, 129, 130, 135, 138

H

Hals, 136
healing, 129, 130, 131, 132, 133, 135, 136, 137, 139
heart disease, 165
heating rate, 62
heavy metal(s), vii, ix, 10, 114, 141, 165, 172, 175, 176
height, 26, 120
heterogeneity, 81
high blood pressure, 165
homogeneity, 58
host, 19, 23, 24, 26, 32, 42
human, 122, 124, 126, 128, 133, 135, 138, 139
hydrogen, 27, 103, 176
hydrogen peroxide, 176
hydrolysis, 2, 12, 20, 27, 41, 156, 166
hydrophilicity, 14
hydrophobicity, 17
hydroxyl groups, 60
hypermineralized, ix, 119

I

illumination, 93, 99
immersion, 94
improvements, 132
in vitro, 124, 126, 137
in vivo, 10, 15, 124, 129, 138
indium, 39
industry(ies), 49, 94, 143, 164, 165, 171
inflammation, ix, 119, 120
inflammatory disease, ix, 119, 120
influenza, 12, 40
influenza virus, 12
inhibition, 39
insertion, 120
integrity, ix, 120
interface, 13, 17, 50, 56, 86, 136
interference, 167
International Atomic Energy Agency, 9
intestine, 20
ionization, 102
ionizing radiation, viii, 92, 93, 102, 104
ions, x, 3, 10, 32, 58, 59, 92, 95, 96, 101, 106, 107, 108, 110, 121, 142, 156
iron, 13, 99, 173, 175
irradiation, viii, 92, 95, 102, 103, 104, 105, 107, 108, 109, 110, 111, 112, 113, 114
irrigation, 125, 127
isolation, 56
isotherms, 145, 149, 150
issues, 173
Italy, 1

J

Japan, 163, 171

K

kerosene, 28
kidney, 9
kinetics, x, 20, 95, 141, 143, 144, 145, 146, 147, 148, 167

L

lanthanide, 35, 42
LDL, 37
leachability, viii, 91
leaching, 176
lead, vii, viii, 45, 92, 93, 107, 108, 109, 110, 111, 112, 113, 114, 173, 175, 176
liberation, 63
ligament, ix, 119, 120, 128, 130, 133, 138
ligand, 7, 8, 143, 151, 152, 156
light, x, 7, 24, 32, 93, 164
light scattering, 7, 24
liquid chromatography, 173
lithium, 43
liver, 8, 15, 39
liver cells, 15
localization, 15
luminescence, 21, 42, 93
lying, 17

M

magnesium, 32, 43
magnetic resonance imaging (MRI), 2, 14, 26, 34
magnitude, 16, 23, 49
majority, 56
management, vii, 165
manganese, x, 142
manufacturing, 164
masking, 165, 171
materials, 14, 32, 38, 46, 47, 50, 58, 93, 94, 98, 115, 142, 151, 153
matrix, ix, 120, 121, 122, 123, 124, 133, 134, 139, 165, 171
measurement(s), viii, 7, 15, 16, 24, 46, 50, 56, 82, 84, 85, 86, 146
media, 12, 50
medical, viii, 91, 92
melt, 31
membranes, 134
mercury, 35, 94
meta-analysis, 134
metal complexes, viii, ix, 10, 14, 32, 58, 59, 92, 94, 98, 100, 101, 114, 165
metal ion(s), ix, 3, 13, 20, 22, 29, 30, 31, 58, 92, 94, 109, 142, 156, 165, 166, 167, 169, 170, 177
metal oxides, 151, 154
metals, x, 2, 13, 38, 92, 94, 104, 106, 108, 109, 110, 142, 143, 148, 151, 155, 156, 164, 165, 166, 170, 175, 176
methanol, 109, 111
mice, 15
microorganisms, 6, 142, 146
microphotographs, 69
microscope, 136
microstructure(s), 49, 50, 51, 53, 54, 69, 81, 84, 85
migration, 120, 121, 124, 126, 128, 130, 132, 142
mimicry, vii, 2, 34
mineralization, 93
miniaturization, viii, 45
Ministry of Education, 141
MMS, 144, 145, 151, 152, 155, 157
modelling, 110
models, 50
modulus, 54
mole, 30
molecular weight, 2, 29, 30
molecules, vii, 2, 20, 26, 34, 60, 100, 102, 103, 104, 142, 166
monolayer, 17
monomers, 23
morphology, 122, 125, 129, 138

mucosa, 135

N

NaCl, 11
nanoparticles, 66
nanotechnology, 14
necrosis, ix, 120, 121, 131
neglect, 51
neutral, ix, 22, 120, 121, 122, 126, 128, 131, 134, 137, 138, 156, 157
nickel, 13, 37
nitrates, 6, 12, 27, 30
nitrite, 109
nitrogen, 16, 95, 114, 165, 166, 169
nitrous oxide, 114
NMR, 8, 23, 24, 34, 39
non-surgical therapy, 132
nuclear magnetic resonance, 35
nucleic acid, 36
nutrient, 165
nutrients, 173

O

OH, ix, 6, 92, 93, 97, 102, 103, 104, 105, 106, 107, 108, 109, 110, 111, 112, 113, 114, 166, 170
OH-groups, 97
oil, 11, 13, 37, 38
oligomers, 19
optimization, 49, 93, 176
organ, 9, 59
organic complexant(s), viii, 91, 92, 93
organic compounds, 19, 63, 94, 99
organic matter, x, 141, 142
organic solvents, 10, 13, 26, 31, 43
oxidation, 93, 96
oxidative stress, 34
oxygen, 32, 49, 95, 104, 109, 114, 166, 176
ozonation, 105, 106
ozone, 176

P

palladium, 35
parallel, 46, 50, 55, 56, 73, 106
patents, 6
pathways, 164
PDL, 138
peptide, 12
periodontal, vii, 120, 122, 126, 128, 129, 130, 131, 132, 133, 134, 135, 137, 138, 173
periodontal disease, ix, 119, 121, 136, 139
periodontitis, 121, 122, 126, 130, 133, 134, 136, 137, 138
permeability, 41
permit, 15
permittivity, viii, 45, 47, 54, 57, 72, 73, 74, 75, 76, 77, 78, 84
phage, 9
pharmaceutical(s), 26, 164
phase transitions, 47
phenol, 94
phosphate(s), 130, 138, 169
photocatalysis, 93, 94, 97, 104, 112
photocatalysts, 94, 97, 98, 114
photocatalytic degradation, viii, 91, 94, 98, 113
photochemical degradation, 105, 174
photodegradation, 93, 94, 95, 97, 98, 99, 101, 106, 114
photolysis, 101, 104
photomicrographs, 125
photons, 93
photooxidation, 100
photosensitivity, 113
physicochemical methods, 169
phytoremediation, 165, 173
placebo, 134
plants, 164, 166, 173
plaque, 121
PM3, 42
polar, 32
polarity, 10
polarization, 47, 86
pollutants, ix, 94, 141, 143

polyamide acid, 31
polycondensation, 31
polyesters, 31
polyether, 32
polymer, 23, 24, 25, 28, 29, 30, 31, 32, 41, 42, 43
polymer chain(s), 32
polymer electrolytes, 31, 43
polymer synthesis, 31
polymeric products, 31
polymerization, 19, 24, 31
polymers, vii, 2, 19, 23, 28, 29, 30, 32, 34, 43
polypropylene, 143, 144
porosity, 97
porphyrins, 41
Portugal, 45, 60
positive correlation, 129
power generation, viii, 91, 92
power plants, 98
precipitation, 31, 58, 60, 106, 107, 112, 156, 170, 176
preparation, 38, 40, 58, 59, 60, 61, 98, 114, 122, 124, 132
probe, 136
prognosis, 120
proliferation, 128, 129, 130, 138
proteins, vii, 2, 4, 12, 133, 135, 139
pulp, 164, 172
purity, 58, 109

R

radiation, v, vii, ix, 91, 92, 93, 94, 102, 103, 104, 105, 106, 107, 108, 109, 111, 112, 113, 114, 115, 175
radicals, ix, 92, 102, 103, 104, 105, 106, 107, 108, 110, 111, 112, 113, 114
radioactive waste, viii, 91, 92
radionuclides, viii, 91, 92
reactants, 60
reaction rate, 103
reaction time, 58
reactions, 2, 58, 59, 93, 102, 103, 107, 109, 156, 165, 166, 176
reactivity, x, 155, 163
reagents, vii, 1, 109, 170, 176
receptors, 38
recession, 131, 132, 134
recognition, 171
recombination, 94
recovery, x, 164, 170, 175, 176
regenerate, ix, 119
regeneration, ix, 119, 120, 121, 124, 126, 128, 129, 130, 133, 134, 135, 139, 176
relaxation, viii, 46, 50, 54, 55, 56, 57, 72, 73, 76, 79, 81, 82, 84, 86, 87
relaxation process(es), 72, 86
remediation, 143, 165, 173, 176
repair, 120, 135, 137
research institutions, viii, 91, 92
researchers, viii, 45, 155, 156
residues, 2, 16, 17, 20, 21, 23, 24, 25, 26, 29, 33, 64, 68, 124, 126
resins, 61, 62, 63, 64, 65
resistance, 30, 40, 52, 53, 54, 55, 56, 73, 75, 77, 81, 85
resolution, 7
response, 49, 50, 54, 55, 56, 57, 79, 81, 120, 138
restoration, 120
risk, x, 164, 175
risk assessment, 175
risks, x, 164, rods, 8
room temperature, vii, 13, 32, 45, 60, 72, 73, 74, 75, 76, 77, 78, 79, 80, 81, 82, 85, 97
root(s), vii, ix, 119, 120, 121, 122, 123, 124, 125, 126, 127, 128, 129, 130, 131, 132, 133, 134, 135, 136, 137, 138, 139
root instrumentation, ix, 119, 121
routes, 53, 164
rutile, 97

S

saliva, 136
salts, 6, 7, 9, 11, 13, 14, 17, 32, 38, 39, 111, 112, 144, 164, 170
saturation, 4, 114

scaling, ix, 120, 123, 128, 130
scanning electron microscopy, viii, 7, 46, 50, 62, 123, 124, 130, 138
scavengers, 106, 107, 109, 110, 111, 114
Schottky barriers, 84
scope, x, 2, 164
sedimentation, 37
sediments, x, 155, 164, 167, 169
segregation, 68, 69, 81
selectivity, 13
self-assembly, 42
SEM micrographs, 66, 67, 71, 72
semicircle, 55, 56, 76, 84
semiconductor, 81, 93, 94, 104
sensing, 35
sensitization, 22
shape, 19, 53, 66, 73
showing, 6, 29
side chain, 14
signs, 61
silver, 72
simulation, 52, 174
single crystals, 49
sintered pellets, viii, 46, 50, 67, 68, 69, 71
sintering, 49, 67, 69, 84, 86
skin, 6
sludge, 93
sodium, 6, 8, 11, 13, 16, 17, 24, 39, 109, 124, 126, 127, 164, 170, 173
sodium hydroxide, 8
sol-gel, 58, 62, 86
solid phase, 104, 108, 112
solid state, viii, 40, 46, 49, 50, 86
solid surfaces, 142
solubility, vii, ix, 1, 10, 18, 25, 26, 34, 92
solvents, 27, 31, 32
sorption, 92, 95, 100
Spain, 1, 34
speciation, 110, 142, 151, 155, 165, 172, 174
species, x, 32, 92, 93, 94, 95, 96, 99, 100, 101, 103, 104, 107, 108, 110, 111, 112, 113, 114, 142, 151, 152, 154, 155, 157, 158, 163, 165, 166, 167, 169, 171, 174
specific adsorption, 155
specific surface, 98, 114
spectrophotometric method, 95
spectroscopy, viii, 8, 32, 46, 50, 53, 67
spin, viii, 46
stability, 6, 7, 11, 15, 35, 37, 39, 48, 58, 59, 101, 165, 166, 168
stabilization, 59, 138
stabilizers, 7
state, 17, 47
steel, 98
sterile, 131
steroids, 38
stimulation, 175
stoichiometry, 19, 23, 26, 58, 68
stomach, 6, 9
storage, 46
stress, 47
stretching, 47
structure, viii, 2, 8, 13, 17, 23, 26, 29, 35, 42, 46, 47, 48, 49, 50, 55, 56, 81, 92, 97, 121, 142, 152
substitution(s), 2, 4, 20
substrate(s), viii, 46, 96, 121, 126
sulfate, 13, 27
surface area, 17, 47, 97, 98, 145, 146
surface modification, ix, 120, 122
surface tension, 6, 7, 10, 13, 16, 18
surfactant, 2, 6, 7, 8, 10, 11, 12, 13, 14, 16, 17, 18, 34, 35, 36, 37, 38, 40
surrogates, 142
Sweden, 117
symbiosis, vii, 1, 34
symbiotic, v, 1
synergic, vii, 1, 13, 34
synthesis, vii, viii, 2, 20, 22, 25, 32, 35, 38, 39, 45, 49, 50, 62, 81, 120

T

tantalum, 46
technetium, 36, 37
techniques, 23, 170, 171
technology, 171
teeth, 120, 122, 128, 130, 131, 133, 135, 136

TEM, 136
temperature, viii, 9, 29, 31, 46, 47, 48, 55, 56, 59, 62, 71, 76, 77, 79, 82, 84, 94, 95, 102, 144
tension(s), 6, 13, 18, 38
tetrahydrofurane, 30
texture, 122, 129, 137
therapy, ix, 2, 119, 120, 132, 133, 134, 139, 173
thermal analysis, 61, 62, 63, 107
thermal decomposition, 59
thermal energy, 93
thermogravimetric analysis, 110
thin films, viii, 43, 46
time use, 130
tissue, ix, 119, 120, 121, 126, 129, 131, 135, 136, 137, 138, 139
titanate, 60
titanium, 94, 95, 98
toluene, 13, 31
tooth, ix, 119, 120, 128, 130, 135, 136
toxic metals, x, 92, 93, 163
toxicity, 39, 94, 107, 170, 171, 175
trace elements, 99
trademarks, 123, 164
transformation, 63, 154, 174
transition metal, 28, 36
transition metal ions, 28
translocation, 173
transport, 38, 95, 142
treated waste, viii, 91, 92
treatment, viii, 8, 9, 27, 49, 62, 71, 91, 92, 93, 101, 121, 124, 128, 129, 130, 132, 133, 134, 136, 139, 140, 164, 166, 170, 171, 173, 174, 175, 176
trial, 134, 139
tricarboxylic acid, 58
trypsin, 122

U

UK, 116, 117
ultrastructure, 137
United States (USA), 117, 166
urban, 143
UV, viii, 8, 91, 93, 100, 105
UV radiation, viii, 91, 93, 100

V

vacuum, 47, 97
valuation, 130, 138
variables, 123
variations, 47, 56
vesicle, 17
viruses, 40

W

waste, viii, ix, 91, 92, 93, 94, 165, 172, 176
waste treatment, 93
waste water, 172
wastewater, 93, 94, 166, 174, 176, 177
water quality, 175
water resources, x, 164
wettability, 6
World Health Organization (WHO), 170, 175
worldwide, 164
wound healing, 120, 124, 133, 135, 136

X

X-ray diffraction (XRD), viii,46, 50, 62, 63, 64, 65, 67, 68, 70, 113

Y

yield, 6, 11, 98, 103

Z

zinc, 13